Water Contamination Emergencies
Can We Cope?

Water Contamination Emergencies
Can We Cope?

Edited by

K.C. Thompson
ALcontrol Laboratories, Rotherham

J. Gray
Drinking Water Inspectorate, London

RS•C

advancing the chemical sciences

The proceedings of the International Conference on Water Contamination Emergencies: Can We Cope? held at Le Meridien Hotel, Kenilworth, UK on 16–19 March 2003.

Special Publication No. 293

ISBN 0-85404-628-3

A catalogue record for this book is available from the British Library

Published by The Royal Society of Chemistry,
Thomas Graham House, Science Park, Milton Road,
Cambridge CB4 0WF, UK

Registered Charity Number 207890

For further information see our web site at www.rsc.org

Printed by Athenaeum Press Ltd, Gateshead, Tyne and Wear, UK

Preface

There is a need to react to emergencies involving low probability/high impact contamination events (chemical, biological or radioactive) in source waters or treated water in time to allow an effective response that will significantly reduce or avoid adverse impacts on consumers or the environment.

The Water Contamination Emergencies: Can we cope? International Conference opened with a detailed consideration of seven diverse examples of significant contamination incidents. These case studies, together with papers from invited experts and poster presentations, formed the basis of a review of lessons learnt and identified ways in which we might better cope in future. The conference was considered appropriate to the time, especially since the conflict in Iraq started in the week of the conference.

Contamination of water can arise from natural causes, industrial accidents, poor watershed management, deficiencies in monitoring or even malicious intent. Examples of incidents attributable to all these aspects, except malicious intent, are been included in this volume.

The key features of an incident are considered to be awareness of the actual occurrence (i.e. "all is not right"); the extent of the incident; accurate and rapid identification and quantification of the chemical/microbiological/radiological species present; risks to the consumer and other water users; the origin of the contamination; short and long term remediation; communications with affected water consumers and other users; liaison with appropriate public health officials and local authorities; and communications with the media and general public. The communication aspect of an incident should never be underestimated Incorrect and alarmist media reports can result in considerable logistic and communication problems for water companies working very hard to contain and remediate the incident

The twenty-one chapters of the book produced by experts in the field of handling water contamination incidents outline various aspects of the above issues. A second and larger conference entitled Water Contamination Emergencies: Enhancing our Capabilities, is to be held in Manchester in June 2005. Full details will be available at http://www.dwi.gov.uk in 2004.

Dr John Gray
Prof. K. Clive Thompson
1 December 2003

Contents

OPENING ADDRESS TO THE WATER CONTAMINATION EMERGENCIES: CAN WE COPE? INTERNATIONAL CONFERENCE, 17 MARCH 2003.

R. Anderson

Department for the Environment, Food and Rural Affairs, Ashdown House, 123 Victoria Street, London SW1E 6DE

1 INTRODUCTION

Looking through your agenda for the next three days and the list of delegates, I had two immediate thoughts. First, it is a real privilege to give this address. The range and depth of expertise represented here today is impressive. And, speaking personally, I must say pretty formidable. Second, the agenda and the range of expertise, brings home the complexity and numerous variety of challenges we face today.

I propose to offer a lay perspective – or, perhaps more accurately, a perspective based on a different kind of experience and responsibility. My concerns are the policies and the framework within which water supply and regulation operates in England. I hope, therefore, that you will see my presentation as contextual.

The title of this conference is Water Contamination Emergencies: Can We Cope? Our ability to cope will, in part, reflect the robustness of our day-to-day arrangements. The robustness of the basic structure for delivery of water services. And, in part, our ability to respond to the unexpected.

2 BASIC STRUCTURE

2.1 How are we doing on the basic structure?

For me, this starts with the standards to which water services are provided. We need to be confident about service levels, about monitoring arrangements, and about enforcement – where necessary. We also need to be realistic about the robustness the system to cope with potential risks.

2.2 Expectations are rightly high.

When I turn on the tap at home and fill my glass with water, I do so in the belief that it is safe to drink it. I make the same assumption here today. The consumer may not know the technical quality standard to which the water must adhere, but assumes that Government will have set out standards and processes that make it safe.

You will not need me to go into the detail of present UK standards for drinking water and for maintaining environmental quality. The key point is that they reflect current scientific advice and are, in almost all respects, common across the EU. The 1980 Drinking Water Directive introduced mandatory standards to protect public health and to ensure that water supplies were wholesome throughout the European Community. Those concerning the environment are set at a level which is proportionate to the risk of pollution posed, while being sufficiently stringent to ensure that there is no observable effect to sensitive aquatic ecosystems.

3 HORIZON SCANNING

It is not enough to rely on what we currently know, we must research and learn. And we must learn to listen. Our record in the UK in the past has not always been successful. We did not fully anticipate the threat from Cryptosporidium. BSE was equally unanticipated. In each case, I think there was a failure to scan the horizon and ask the right people the right questions.

We must engage with those at the fringes of the Department's activities, including those who might have conflicting objectives or opinions. Not just those at the heart of what we do. And not just those who agree with us. And this needs to be built into the research process.

With this in mind, my Department has now incorporated "Horizon Scanning" as a keystone of its science strategy. An open house invitation was extended to individuals and organisations to comment on:

- Potential threats and opportunities, including socio-economic aspects;
- Vulnerabilities of sectors for which DEFRA has responsibility with a view to developing policies that reinforce resilience; and
- Novel or critical perspectives on current policies with a view to improving the robustness of the policy making process.

As a result, across the Department we are focussing on five priority themes:

- Future landscapes and the influence of land use and environmental planning;
- Environmental constraints and the impact of the reducing availability of natural resources;
- Re-thinking the food economy;
- Coping with threats – society's ability to adapt to environmental change and the impact of new diseases; and
- Meeting people's future needs.

These themes go much wider than water. But, by taking this broad interlocking approach, we seek to ensure that research on water issues is seen in a wider context.

4 RESEARCH EFFORTS

So, looking forward, where should we concentrate our research efforts? A substantial proportion of the drinking water research budget is likely to be devoted to characterising the possibility of risk amplification arising from the coincidence of different low frequency / low risk events. By way of example I will mention two areas.

One is storage of water for drinking within buildings. This might create conditions that favour the multiplication of bacterial species that would not normally present a risk via drinking water.

Another is low concentrations of disinfection by-products in water. Showering or swimming could give rise inhalation of volatile DBP and could present amplified risk at certain stages of pregnancy.

In the field of environmental water protection, the focus of current research is towards understanding and addressing challenges such as agricultural and urban diffuse pollution and climate change and towards underpinning the regulatory regimes relating, for instance, to dangerous substances.

As we learn more from the results of future research, the findings can be fed into future revisions of standards. As we saw with the way in which the Drinking Water Directive was revised in 1998 to take account of the updated World Health Organisation guidance on water quality parameters.

The new regulations introduce a number of new standards and more stringent values for some existing parameters. The most important change is for lead where the current standard for lead of 50 micrograms per litre is tightened to 25 micrograms per litre by the end of 2003 and 10 micrograms per litre by the end of 2013.

5 MINIMISING RISK

So we have mechanisms for setting clear standards. We have processes in place to explore and assess potential threats. How then to ensure that the standards are delivered and day-to-day risks minimised? There are a number of partners and all have vital roles:

- There are the water companies, have the legal responsibility for the wholesomeness of public water supplies;
- There is the Drinking Water Inspectorate. The main role of the Drinking Water Inspectorate is to ensure that Water Companies supply water which complies with the standards set down. This it does mainly through regular audit of compliance data and by regular inspection of the Companies. Any deficiencies identified are required to be rectified by the Company;
- Responsibility for monitoring the wholesomeness of private water supplies rests with Local Authorities. Local authorities have legal obligations to inspect and take action when necessary; and
- The Environment Agency monitors and enforces environmental water standards.

6 DRINKING WATER INCIDENTS

When an event affects, or is likely to affect, the wholesomeness or sufficiency of drinking water supplies the Drinking Water Inspectorate is the focal point for receiving notifications made by water companies. It assesses the information provided and the actions taken by the company. If there were contraventions of standards, a decision is taken as to whether initiate enforcement action and, if water unfit for human consumption was supplied, whether a prosecution should be initiated.

In an emergency, the Inspectorate would be involved in providing technical and scientific advice to those managing the emergency. This would include advice to my Department and to Ministers.

Although we tend to take for granted clean and wholesome drinking water, this remains on of my Department's key objectives. There is no room for complacency. It is vital that we retain the confidence of consumers in the safety of drinking water supplies by maintaining high standards.

7 ENVIRONMENTAL WATER QUALITY

Turning from drinking water to environmental quality standards, the Environment Agency is the main organisation responsible for protecting and improving the environment – including the water environment – in England and Wales. The Agency's objectives include:

- tackling flooding and pollution incidents;
- reducing the impacts on the environment from industry and agriculture;
- cleaning up rivers, coastal waters and contaminated land; and
- improving wildlife habitats.

The way in which environmental quality standards for water are applied is crucial to their effectiveness, and the associated compliance regime must be appropriate.   In England and Wales, environmental quality standards for surface waters are used by the Environment Agency to calculate emission standards in conditions in discharge consents, as well as to assess the quality of the aquatic environment.

It is worth adding that a range of approaches is possible when applying standards. Typically in the UK, they are determined by obligations under European and UK legislation and expressed in Regulations.   This is particularly true of aquatic standards.   However, there is also role for more flexible measures, such as general binding rules, codes of good practice or voluntary agreements.

The application of environmental quality standards for the aquatic environment has played a crucial role in improving the chemical quality of the aquatic environment in England and Wales over the past 10 years.

The work carried out over the past 20 years on aquatic environmental quality standards and their related compliance regimes has provided regulators here with a far better understanding of the impacts of risks posed by pollutants and has enabled the UK to contribute positively to discussions on standards at a European level.   An example is the Water Framework Directive.

As with drinking water, we also need to ensure that we are looking ahead. New issues – such as the potential effects of endocrine disrupting substances –need to be properly studied and understood.

8 EMERGENCIES

Returning to the theme of this conference - can we cope?   We can set standards. We can set up processes for enforcement. We can seek to anticipate future threats. But, we will not always be successful. If contamination does occur, how resilient is the system? Will consumers still be able to get basic services? Will the environment continue to be protected?

For drinking water, each water company is required to have emergency arrangements in place to cope with what they have judged to be their worst-case scenario. These plans are reviewed annually. They include measures enabling one company call upon the help of another – mutual aid. We have, as you might expect, reviewed these arrangements following 11 September.

We have also been providing a series of briefings on security matters to boards and senior managers of most water companies as well as Scottish Water and OFWAT. We have worked with the industry and the Environment Agency and Scottish Environment Protection Agency to develop a Protocol for the Disposal of Contaminated Water. Work has been started to identify suitable methods of treatment of contaminated water, to permit subsequent safe disposal.

In the last resort, if drinking water is contaminated arrangements exist to ensure that there is an alternative supply.

Where contamination affects environmental quality, the option of alternative supply is rarely available. Instead, the focus is, rightly, upon how the effect can be minimised or contained.

The Environment Agency has long-established procedures for dealing with contamination and potential contamination to ground and surface waters. These procedures are designed to ensure there is effective control and co-operation between the emergency services, local authorities and water industry.

9 CONCLUSION

To sum up, we have, in England and Wales, seen a steady increase in the quality of drinking water and in the aquatic environment. In 2001, 99.86% of the 2.8 million tests showed compliance with the standards. Breaches of the standard were one twelfth of what they were in 1992. There were no reported confirmed cases of cryptosporidiosis related to the mains drinking water supply. 98% of coastal bathing waters met the European Bathing Water Directive and 94% of rivers were of good or fair chemical and biological quality. There is much to protect.

But things can go horribly wrong. Past success is no guarantee of success in the future and where mistakes have been made in the past we must learn from them. Contingency Planning has always been important. It has never been more so. Ministers have made it very clear to water companies and the regulators that they expect robust plans to be in place. An aim that I am sure that you will share. Your conference could not be more timely. I hope that it will help address issues such as where are the gaps? What further needs to be done? Are the lead responsibilities clear? And how can we learn from each other?

Unfortunately, I will not be able to stay for the full term, but my colleagues Nigel Cartwright and John Gray will be here. I will be interested in their report back. The next three days represents a rare opportunity to compare notes; to find the read across from different situations and perspectives; and to look beyond the immediate.

I would not be surprised if it was concluded that aspects of our present systems do warrant further scrutiny. We must be ready to respond so that water supplies continue to meet the standards consumers expect, the environment is protected, and we have a system which is resilient to attack.

Thank you. Have a good conference.

THE CITY OF MILWAUKEE *CRYPTOSPORIDIUM* OUTBREAK: WHAT REALLY HAPPENED AND HOW VULNERABLE ARE WE?

D E Huffman

College of Marine Science, University of South Florida, 140 7th Ave. South
St. Petersburg, Florida, USA

1 INTRODUCTION

It has been 10 years since the City of Milwaukee, Wisconsin's drinking water became contaminated with the protozoan parasite *Cryptosporidium parvum*. This single drinking water outbreak would become the focal point of hundreds of research studies focusing on the cause of this outbreak as well as methods for the prevention of future outbreaks of this magnitude. Could this outbreak have been prevented and what have we learned over the last decade with regard to *Cryptosporidium parvum*?

2 *CRYPTOSPORIDIUM*

If we begin with the basic biology of this organism, *Cryptosporidium parvum* belongs to the phylum Apicomplexa (referred to as the Sporozoa) and is one of approximately 5,000 species. It is a coccidian parasite that infects the gastrointestinal tract and has become one of the most important enteric pathogens in both humans and animals. The parasite was first described in 1907 by Tyzzer[1], but over the last two decades there has been an escalation of published material on the biology, genomic characterization, taxonomy, transmission, detection and public health risks associated with this organism.[2-6] There are currently ten recognized species within the genus *Cryptosporidium*, namely, *C. baileyi* and *C. meleagridis* found in birds, *C. felis* found in cats, *C. muris* found predominately in mice, and *C. wrairi* in guinea pigs, *C. andersoni* in cattle, *C. nasorum* found in fish, *C. serpentis* found in reptiles and *C. saurophilum* in skinks, but the species of concern from both a medical and veterinary perspective is *C. parvum* (2). Current research has identified two different genotypes of *Cryptosporidium parvum* that are infectious in humans.[7] Genotype 1 isolates, which have only been shown to be infectious in humans and genotype 2 isolates which have been shown to be infectious in mice, calves, lambs, goats, horses as well as humans. This suggests the possibility that there are two distinct populations of oocysts cycling in humans with distinct transmission cycles; 1) zoonotic transmission from animal to human with subsequent human to human and human to animal transmission and 2) a transmission cycle exclusively in humans. While the number of species and genotypes of *Cryptosporidium* may seem highly technical and unrelated to the question of what really happened and the vulnerability of Milwaukee, it will become clear that determining the source of the contamination would provide information to assist in the determination of system vulnerability.

3 WATER SUPPLY TO MILMAUKEE

Two drinking water treatment plants served the City of Milwaukee. The Linwood Plant, put into service in 1939, primarily served the northern two thirds of the city while the Howard Avenue water treatment plant (HWTP), put into service in 1962, served the southern one third of the city. Both treatment plants received their raw water from Lake Michigan. Potassium permanganate feeders were placed at the Texas Ave. Pump Station that pumped water from Lake Michigan 3.5 miles to the HWTP. Both treatment plants practiced conventional water treatment processes (coagulation, sedimentation, filtration and disinfection). Alum sulfate (alum) was used as the coagulant at the HWTP until August 1992 when the facility decided to switch to polyaluminum chloride (PACl). The switch to PACl was made for several reasons: 1) to meet the requirements of the United States Environmental Protection Agency (USEPA) lead and copper rule 2) to reduce the sludge volume and 3) for improved coagulation of raw water under cold conditions. [8] The floc produced by PACl is smaller and denser than that generally produced using alum and uses charge neutralization as the primary mechanism of particle destabilization. PACl also increases finished water pH thereby reducing the level of lead and/or copper leaching from residential plumbing. Initial lead/copper sampling in June 1992, revealed that the HWTP while in compliance could benefit through the change of coagulant to PACl however, a second round of sampling in January 1993 after switching to PACl (September 1992) revealed the HWTP was out of compliance with the lead and copper rule. The HWTP continued to perform jar testing with PACl to optimize their water treatment capability.[8]

4 THE INCIDENT

The early spring of 1993 in Michigan was marked by a series of heavier than normal rainfall. In the weeks leading up to the Milwaukee boil water advisory, increasing amounts of run off from agricultural areas as well as storm water and sewage catchments were diverted into Lake Michigan. Turbidity levels in the raw water intake for the HWTP fluctuated dramatically with a low of 1.5 to a high of 44 NTU.[8] Other drinking water treatment plants whose intakes were located along Lake Michigan also noted turbidity spikes in the raw water. The total coliform levels in the raw water were also markedly variable ranging from <1 CFU/mL to approximately 3,200 CFU/mL.[8] During the weeks of March 24 through April 2, the PACl dosage was radically adjusted on an almost daily basis to combat the high levels of turbidity in the final effluent (1.1 to 1.7 NTU).[8]

Difficulties in turbidity reductions were exacerbated by a marked increase in customer complaints regarding taste and odor. The treatment plant operators responded to complaints by adjusting the levels of powdered activated carbon (PAC) and potassium permanganate as well as PACl that were being utilized. Treatment plant operators along with the facilities chief chemist and representatives from the PACl supply company worked day and night to try to ascertain the optimum coagulant dosage. The fluctuating turbidity in the raw water combined with the short residence time in the plant along with recycling of the filter backwash water to the head of the plant, combined to exacerbate the situation. Throughout this time period, combined filter effluent turbidity and total coliform levels did not exceed the USEPA Surface Water Treatment Rule. While filter effluent turbidimeters were in place within the treatment plant they were not operational at the time of the outbreak.[8] It has been speculated that a lack of historical data on the use of PACl at the HWTP contributed to the inability of the facility to determine optimum dosage levels. The HWTP decided to switch its coagulant back to alum on April 2nd and over the next few days, turbidity levels in the HWTP final effluent were reduced to near normal levels (0.5 to 0.15 NTU). It was not until April 7th that the Wisconsin state epidemiologist noted the presence of *Cryptosporidium* oocysts in stool samples from three Milwaukee area patients with gastrointestinal illness and watery diarrhea. A concurrent survey of diarrhea in local nursing homes revealed that nursing home residents located in the southern portion of the city (served by the HWTP) were 14 times more likely to have had diarrhea than residents in the northern part of the city. The Mayor issued an advisory to Milwaukee Water Works customers to boil their water and the HWTP was ordered to be closed on April 8th.[8]

The HWTP became an icon for system vulnerability. A series of events including climatic change, aging infrastructure and a lack of adequate watershed protection all contributed in some part to the Milwaukee *Cryptosporidium* outbreak. While it may be impossible to control climatic conditions, watershed protection within the range of climate extremes should be explored. The incidence of infection in the animal or human population and the excretion of oocysts in watersheds are important factors contributing to waterborne disease. The size of the watershed, dilution and hydrology of the system as well as the type and reliability of the drinking water treatment system will influence the impact of pathogens in the drinking water.

Deliberate introduction of a disinfection resistant microbe into a water system could result in illness. While the amount of material needed to deliberately contaminate water sources (such as a reservoir or aquifer) is large, contamination of a public water supply does occur accidentally (*Crypto/Giardia/E.coli*). Early finger pointing implicated cattle manure spread fields in the watershed as the source of *Cryptosporidium* oocyst contamination. However, recent advances in molecular methods have classified oocysts from four persons infected during the outbreak to be identified genetically to be of human origin with the probable source being sewage overflow in the Milwaukee watershed.[7]

5 COSTS OF THE INCIDENT

Improvements to the Howard Avenue Plant included relocation of the raw water intake by 4,200 feet at a cost of $11 million dollars.[9] Land use restrictions as well as specific requirements for household waste and combined sewage overflows that previously emptied into the Milwaukee Harbor have also been delineated. Upgrades to the water treatment plant include optimization of filtration by improvements to the chemical feed systems, new filter media (anthracite over sand) supervisory control data acquisition (SCADA) real time information on pressure, flow, power, chemical dosages, residuals and system status which is automatically controlled for plant pump rate changes. Finally, both the Linwood and the Howard Avenue plant were retrofitted with ozone treatment in July 1998, at a cost of $38 million dollars. Ozone can be utilized for organics oxidation, pesticide oxidation (AOP), coagulation assistance, promotion of biofiltration, control of disinfection by-product precursors and finally to lower chlorine demand. The USEPA has estimated that $138.4 billion dollars will be necessary to replace or upgrade the US drinking water infrastructure over the next 20 years.[10]

The value of preventing an outbreak of cryptosporidiosis can be seen in medical expenses, lost wages, lost leisure time, changes in life expectancy and risk of premature death. These costs must be balanced against changes in land use, source water, water treatment and the effect of issuing a boil water order. The cost associated with a mild case of diarrhea has been estimated at approximately $280 per person (lost work, medicines) with a severe case of diarrhea costing almost $8,000 per person for medical diagnosis, treatment etc. The costs associated with the Milwaukee outbreak, with an estimated 400,000 cases of cryptosporidiosis, have been estimated at greater than $55 million dollars. The Centers for Disease Control and Prevention estimated that during the Milwaukee outbreak, only 1 out of 23,750 cases of cryptosporidiosis were detected before there was public recognition that there was a waterborne outbreak.[10] This may have been due to the lack of expertise in the identification of this organism in the clinical laboratory or simply the inability of the clinician to order the proper parasitic examination.

6 CONCLUSIONS

Disease surveillance and investigation can provide a crucial link in the prevention of waterborne disease. Increased physician awareness of infectious waterborne organisms such as *Cryptosporidium* as a cause of watery diarrhea along with more sensitive clinical laboratory methods have increased the identification of this parasite within the population. The safety of drinking water supplies cannot be taken for granted. Outbreaks of waterborne diseases are continuing to be documented worldwide and it is only through a commitment to improved vigilance in surveillance, along with improvements in science and technology that we can hope to maintain safe drinking water in the 21st century.

Acknowledgements:

The author wishes to thank James E. Christopher, P.E. of Hartman and Associates Inc. for information on Howard Ave. Plant during the months leading up to the Milwaukee outbreak.

References:

1   E.E. Tyzer, Proc. Soc. *Exp. Biol. Med.*, 1907, **5**, 12.
2   R. Fayer, U. Morgan, and S.J. Upton. *Int. J. Parasit.*, 2000, **30**, 1305.
3   H.V. Smith and J.B. Rose, *Parasitol Today*, 1998, **14**, 14.
4   J.B. Rose, *Ann. Rev. Pub. Hlth.*, 1997, **18,** 135.
5   W. Quintero-Betancourt, E. Peele and J.B. Rose, *J. Microb. Methods* 2002.
6   A. Wiedenmann, P. Krüger and K. Botzenhart, *J. Indust. Microbiol. Biotechnol.*, 1998, **21**, 150.
7   M.M. Peng, L. Xiao and A.R. Freeman *Emerg. Infect. Dis.,*1997, **3**: 567.
8   K.R. Fox. and D.A. Lytle, 1996, *J. AWWA* **9**, 87.
9   K. Blair, 1995, *J. Env. Hlth.* **58**, 34.
10 USEPA Drinking Water Infrastructure Need Survey: First Report to Congress. Washington, D.C. January 1997.
11 Juranek and Levy, US EPA Office of Ground Water and Drinking Water. Stage 2 Microbial/Disinfection Byproducts Health Effects Workshop, Washington, D.C. 1999.

THE WEM INCIDENT

M. Furness

Quality and Environmental Assurance, Severn Trent Water

1 COURSE OF EVENTS

On the morning of Friday 15th April 1994, Severn Trent Water experienced a drinking water quality problem which came without warning, originated 120km away, passed through a treatment works, inconvenienced and alarmed 110,000 customers bringing local and national attention. The works in question employed conventional clarification, filtration and disinfection treatment. However, unlike other company works, activated carbon had not been introduced because the works was to be closed.

Customer complaints of an unusual taste and odour in the public water supply to Worcester started to come in from 0750 hrs on 15 April. It was evident to the Customer Services Bureau that these were consistent with a problem at Barbourne Water Treatment Works on the River Severn. The Works Operator immediately confirmed by his own tests that there was an odour and taste problem in the final water. He immediately took steps to shut down the pumps.

The Worcester and District Health Authority was informed by the Severn Trent Quality Assurance Department that there were some 15 complaints of abnormal odour and taste in the Worcester area. An Emergency Team was set up at Company Headquarters and the process of the provision of alternative supply of water by tankers began.

Later in the morning the number of complaints had risen to 60 and the decision, in collaboration with the Consultant in Communicable disease Control (CCDC), Worcestershire H.A, was made to advise the customers in the Worcester area not to drink the water. Note that the health risk assessment was made even more difficult by the lack of consistency over the odour description, e.g., sweet, sewage, paint stripper etc.

At 1230 hrs a meeting took place of the Worcester Health Emergency Incident Team (HEIT) comprising of representatives from the District Health Authority, Severn Trent Water, Officers from Wychavon, Worcester City and Malvern Hill District Councils, County Scientific Services and the National Poisons Centre, Birmingham. The Team were given regular information by Severn Trent on the areas affected. A helpline was made available at the Worcester and District Health Authority to give health advice. Press statements, together with media interviews were also given.

A Crisis Management Team, chaired by the Severn Trent Managing Director was convened at Severn Trent Water and worked in close association with the Worcester Emergency Incident Team. In the meantime, local media was helping by broadcasting warnings not to drink the water and the process of notifying hospitals, schools and business customers was underway.

The first press statement issued by the Health Emergency Incident Team (HEIT) advised customers that an organic chemical had entered the water supply system and first indications were that this did not pose any serious threat to health. Until further information became available, the public were advised not to drink the water or to use it in food preparation.

Strensham and Mythe Water Treatment Works, further downstream on the R. Severn were shut down as a result of taste and odour tests. The affected water had not penetrated the Mythe Works but had done so at Strensham WTW which supplies Coventry. The flow up aqueduct to Coventry was reversed quickly enough for Warwickshire and East Worcestershire not to be affected. The NRA also reported that the river was tainted north of Bewdley.

A letter drop to 30,000 customers of Severn Trent in the Worcester area began. The letters were printed during the afternoon in Birmingham, delivered to Worcester and distributed by local members of staff. At about the same time, local TV interviews were given advising of the extent of the problem and warning customers that Gloucester, Birmingham and Wolverhampton might be affected (the latter via the South Staffordshire Hampton Loade works). Appeals were made to customers in Coventry and the East Midlands to conserve supplies because the Strensham Aqueduct had been closed as a precautionary measure.

Requests were made for additional tankers and bowsers from Thames Water, Welsh Water and Wessex Water. At 2005 hrs the West Midlands Regional Health Authority advised that Worcestershire Water was safe for bathing.

Early on Saturday 16th April, there was a report of a strong solvent smell coming into Wem Sewage Works in Shropshire. The location of the discharge was identified at Wem Industrial Estate and the National Rivers Authority (NRA) served a formal sample on Severn Trent Water at Wem Sewage Works. The NRA and Police attended the site of the suspected discharge to the sewer and samples were taken for analysis. The decision was taken to intercept the material at the pumping station and tanker it out of the catchment. Legal samples were taken by Severn Trent and the NRA of the suspected discharge and the company believed to be involved were informed.

By 0750 hrs there was the indication of the presence in the river and treated water samples of two chemicals, also seen in the discharge sample, namely 2 EMD (2-ethyl-4-methyl –1, 3–dioxolane) and an unknown substance. Both chemicals were at sub ppb levels in the potable water supply.

Early in the morning the Crisis Management Team met with the Regional Director of Public Health, West Midlands Regional Health Authority. He agreed that the water might taste and smell unpleasant for the next couple of days, but that there was no threat to public health.

On Sunday 17th April, there was a rapid improvement in the situation. By 1235 hrs the NRA reported that the River Severn had no remaining trace of odour. Customer complaints rapidly subsided. The bowsers were progressively withdrawn.

2 THE CHALLENGES

2.1 Alternative supplies

Arrangements for the provision of bowsers, mostly 250 to 500 gallon capacity, in Worcester and the outlying affected areas were put in hand. Altogether, 250 bowsers were placed at strategic points in the streets and a large number of tankers together with their drivers were brought into use to fill and replenish the bowsers with treated water from unaffected sources within Severn Trent area. Valuable assistance was provided by Thames Water, Welsh Water and Wessex Water and other owners of suitable tankers. The process of Mutual Aid had worked very effectively.

2.2 Information clamour

From the early hours of day one the demand for information grew in intensity and quickly became an exhaustive process. As well as Severn Trent's three emergency teams and supporting groups there was an obvious need to communicate with:-

The public
The Regulators
The Health Team
MPs and of course the Media

There is no doubt that the Media were vital in their ability to warn the public via local and national radio and television, and in this supporting role were invaluable. However, the printed word did not always reflect the same picture, especially as the days passed into weeks and even many months later.

The following day the local paper reflected that the "Water woe to continue". It also gave the first hints of the need for compensation and investigations. The paper further described the shoppers stampede and 'where were the alarm bells' was reflected in the Birmingham paper. Nottingham papers talked about a Government inquiry, while a Manchester paper reflected the story of thieves stealing water tanks. By Sunday, the major national papers were involved. The Times asked "why the water company failed to detect the chemical spill", while The Independent was headlining a likely court case. The local Sunday paper issued a warning that it could happen again and targeted the tanker thieves again.

Stories continued to appear for a further 11 days.

2.3 Odiferous compounds

It has always been known that certain chemical compounds can impose a highly detectable odour even in trace quantities. The potential impact on potable water systems was thought to be negligible. There was no archive reference available on such compounds. It was believed that we would be able to identify them through conventional laboratory analysis.

The potent odour produce in this incident 2EDD (2-ethyl–5,5-dimethyl-1,3-dioxane) possessed an analytical anomaly that meant it could not be identified by conventional gas chromatography mass spectrometry (GCMS). Indeed it was a further 13 days before it was identified.

2.4 Passage through treatment

Chemical products passing from river sources through water treatment are traditionally removed as part of the process. The organic compounds involved in this incident were not broken down through coagulation, filtration and chlorination. Absorption through carbon treatment can be effective. There was no advanced carbon treatment at the works as Barbourne Treatment Works was to be shut down. Carbon treatment was already on the investment plan for the other direct river abstraction works.

2.5 Health risk assessment

The public health advice given verbally and early to Severn Trent was that people should be advised not to consume the water. This was confirmed when the Health Emergency Incident Team met later at lunchtime on the Friday. Potentially vulnerable groups such as nursing homes, the Worcester Royal Infirmary, nurseries, plan groups, schools and residential homes were all alerted through pre-arranged cascade systems as were all food producers. GP's were contacted through a telephone cascade system.

The HEIT were given information by Severn Trent on the geographical areas affected and on water analysis results as the situation developed. A helpline was set up at the Health Authority to give health advice. Press statements and media interviews were organised as new information became available.

One of the problems facing the HEIT was the variable nature of the complaints which Severn Trent was receiving from customers. The smell was described variously as sewage, bad eggs, paint stripper or paint. The first press statement issued at 3.00pm advised the public that an organic chemical had entered the system and that preliminary checks indicated that it did not pose any serious threat to health. As a precaution, in the absence of any further information, the public were advised not to drink the water or use it in food preparation.

There were also difficulties in saying whether people could use the water for washing but, given the low concentrations, it was eventually decided that it could be.

Staff from all agencies involved in the Team helped to man the eight helplines from 3.00 pm on Friday. The helplines remained operative until Monday morning 18th April and handled an estimated 3,500 calls.

At 1.00am on Saturday 16th April, news was received that the source had been traced to a solvent recovery company in Wem, Shropshire and further discharge to the river had been stopped. It was estimated to take 36 hours for the contaminated water to clear Worcestershire.

Mid-afternoon on Saturday 16th April the two chemicals contaminating the tap water were identified. These being 2-ethyl-4-methyl-1,3-dioxolane and 2-ethyl–5,5-dimethyl-l,3-dioxane which was causing the unpleasant odour and taste. Both were found in concentrations of less than one part per billion and medical toxicologists from the National Poisons Centre and the Department of Health advised that these chemicals at the very low concentrations found were not hazardous to health. A press statement was issued at 5.30pm assuring the public that the water was safe to drink although taste and smell problems may still occur in some areas.

3   AFTERMATH

Severn Trent Water undertook a comprehensive review of all its operational and response procedures associated with this incident. A national seminar was organised for the Water Industry.

An independent enquiry was undertaken by a team led by Professor Ken Ives and a report duly published with 22 key recommendations.

Compensation claims did arise, particularly from businesses who claimed to have tainted products.

The local Health Authority conducted an immediate epidemiological survey, followed by a retrospective cohort study several years later. The results did not show any cause for concern with respect to the safety of the drinking water.

4   IMPLICATIONS FOR THE WATER INDUSTRY

Severn Trent Water was prosecuted for the above incident for "supplying water not fit for consumption despite the fact that there was no risk to health and complimentary comments from the Judge on the effectiveness of the emergency response.

There were a number of recommendations from the independent report which were aimed at the Water Industry as a whole. Further review within the Water Industry had led to the following improved procedures:-

- More frequent on site taste/odour testing at treatment works.
- Carbon treatment for direct river abstraction supplies
- A toxicity database which include a list of odiferous compounds.
- A formalised Mutual Aid group (national)

In reaching the final decision in court, the Judge decided that if customers did not like the taste then the water was unfit. This ruling previously had only been applied where there were health implications. The ruling of unfit has been applied to water quality incidents since Wem.

In summary, this was a unique incident which although well handled by the Water Company and other agencies at the time did lead to a number of key lessons for the Water Industry which should ensure a similar incident could not be repeated.

THE HUNGERFORD FISH MORTALITY

G.C. Brighty

Environment Agency, Evenlode House, Howbery Park, Wallingford, Oxon , OX10 8BD
UK

1 INTRODUCTION

This paper differs from most of those presented at this meeting in that it describes an impact in the environment that had all the hallmarks of a typical pollution incident but, strangely, without an obvious cause. This of course leads to difficulties in decision-making because we are faced with many uncertainties. The ramifications of not knowing what is behind a large environmental impact are far-reaching, for water users, for those whose businesses depend on the watercourses and the general public. The media also played their role, putting the incident and those involved under the spotlight. For the Environment Agency and British Waterways, the two organisations primarily responsible for the waterways in question, the Hungerford fish mortality was a rigorous test of the public's confidence in our abilities to manage incidents, investigate the causes and clean up the environment.

Although this was primarily an environmental incident, considering the implications for human health *via* the environment becomes a very high priority, and setting precautionary action in motion an imperative. The Hungerford Fish Mortality also was a dynamic event, in that the severity of the effects seen on the fish environment varied over time in the watercourses, which is unusual. If this event taught us one thing, it is always to look at all the information available with an open mind.

2 OVERVIEW OF THE INCIDENT AND IMPACTS

2.1 Location and extent of the incident

The incident was focused on watercourses close to the Berkshire town of Hungerford (Figures 1, 2). The interconnecting nature of the two streams, river and a canal, and roads and railway lines at this point were major complicating factors both for the investigation and management of the incident.

An Atlantic storm passed over the British Isles on the night of $3^{rd}/4^{th}$ March 1998, bringing with it the first significant rainfall for several weeks. In the early hours of 4th March, The Environment Agency's 24-hour communication centre in Reading received a call from the Berkshire Trout Farm at Hungerford, reporting tens of thousands of dead and dying fish. An Environment Protection officer was immediately sent to investigate,

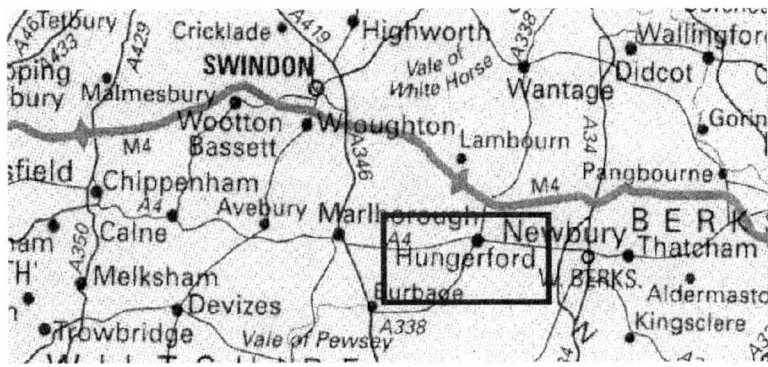

Figure 1   *Location of the Hungerford Fish Mortality, Hungerford, Berkshire UK*

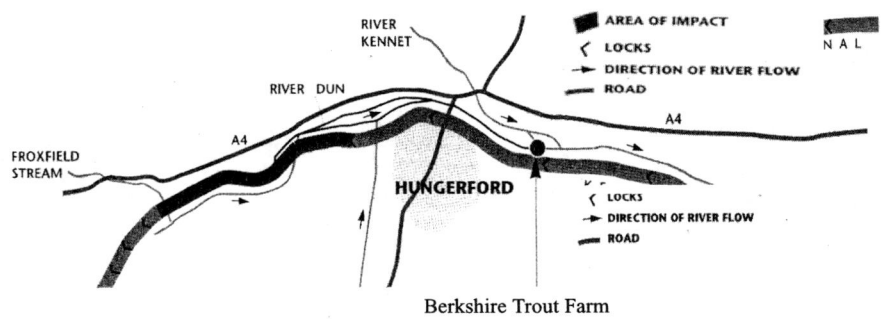

Berkshire Trout Farm

Figure 2   *Schematic diagram of the watercourses and areas of impact*

beginning with the trout farm and surrounding waters, the Kennet and Avon Canal, the River Dun and Froxfield Stream, taking measurements of water quality and collecting samples for later analysis. Having confirmed the serious nature of the incident, the Agency's Thames region emergency plan was put into action, setting up an Incident Control Room at Wallingford to coordinate the operation.

By dawn, a clearer picture was emerging. The total stock of the fish farm had been affected with many thousands of trout already dead and the remainder sluggish and unresponsive. It soon became clear that the incident had originated in the Kennet and Avon Canal, near Froxfield, and in the River Dun had suffered the same fate. The pollution in the River Dun, which feed the Berkshire Trout Farm, had then passed into the farm with disastrous consequences. However, although many fish had died, no other forms of wildlife – invertebrates, birds or mammals – had been affected. Also, no fish were dying downstream of the trout farm, indicating that the farm had effectively stemmed the movement of the toxic material.

2.2 Initial investigations and clean up

The results of the chemical analyses performed on the first water samples taken came through just before midday on the 4th March from the Agency's laboratory in reading. No traces of cyanide or agricultural pesticides – common causes of fish kills – had been found. All of the Agency's routine water quality indicators such as dissolved oxygen and ammonia, were giving normal readings. Samples of fish from the farm and the wild were collected for detailed analysis including blood screening, tissue analysis and detailed organ examinations

Meanwhile, the Agency field staff were systematically searching the area surrounding the farm, defining the extent of the impact and looking for clues as to the possible cause. The local fire service were asked for information on any incidents that they may have attended, Railtrack (who manage the railway infrastructure) were contacted for any unusual events on the railways or operational activity such as spraying of the tracks and local farmers were asked for their crop spraying records. British Waterways were also involved because, as navigation authority for the Kennet and Avon canal, they had been carrying out dredging. Thames Valley Police flew their helicopter over the affected areas, taking video and thermal imaging data.

There were concerns that the toxic substance, whatever it was, might pass downstream and pollute the canal, the nearby River Kennet with its high quality fishery and potentially cause a risk to water abstractions for human use and other fish farms further downstream. As a precautionary measure, the Agency advised British Waterways to close the canal in the affected area. This prevented the use of the canal for navigation, for boats and canoes and the public were advised not to go to the waterside in case they too put themselves at risk. Fishermen were similarly advised not to fish on the system. Daily briefings were given to broadcast media and local stakeholders, including public health. It was a big incident attracting much interest.

Figure 3   *Impact on Berkshire Trout Farm*

The clean up began on 5 March with the first removals of the 500,000 dead fish taken from the farm ponds and river system. Because the nature of the pollutant was unknown, the fish had to be classified as hazardous waste and taken in skips to a licensed landfill, in a clean up operation that took 5 weeks. In total, 150 tonnes of fish, mostly comprising the market-ready fish from the Berkshire Trout Farm, were removed.

The post-mortem examinations of the fish revealed severe damage to the gills – swelling or hyperplasia - the worst ever seen by the fish pathologist. Fish had been seen swimming erratically and "gasping" at the surface. Some internal damage, including haemorrhaging from the kidneys was evident, but the conclusion was that the fish had suffocated, despite there being sufficient oxygen in the water. There were some clues as to what was the effect, but little to go on for the causes or their source into the canal system. Rumours were abounding locally about sabotage of the farm, but there was no "smoking gun" to support any of these. After 5 days of the incident, it was time for a re-think.

3   SCIENTIFIC INVESTIGATION

3.1   Review of the evidence

3.1.1   Chemistry   The initial and subsequent chemical analyses of water, sediment and fish tissue had not revealed any man-made pollutant at a concentration likely to cause death to fish. All the datasets collated for the substances either "best guess" or from broad-scan chemical analyses were reviewed by the Water Research Centre (WRC). Although the data were satisfactorily reviewed, further candidates were suggested based on reviews of pollutants known to be particularly toxic to fish, and less so to invertebrates. A further 60+ substances were identified including organophosphorus and synthetic pyrethroid pesticides, but of these none were found at even approaching LC_{50} values. With little else to consider, chemistry was put on hold until new investigation leads emerged.

3.1.2   Algology Algal samples previously taken for identification of major taxonomic group composition and count estimates were re-examined for known toxic algal species, such as *Oscillatoria* and *Microsystis*. Water samples were retaken and similarly examined, and also tested for mammalian toxicity. No toxic species were found and no samples proved to be hepato or neuro-toxic, effectively ruling out the involvement of any cyanobacterial species.

3.1.3   Site investigations   In parallel to the local site visits, biologists examining the extent of the impact documented information on the state of the environment in the vicinity of the canal. It was noted that in the preceding period in February, large algal blooms comprising mostly diatoms dominated by *Stephanodiscus sp.*, had occurred in the canal. This would be a relatively unusual phenomenon at that time of the year but February 1998 had been an unusual month, with clear skies, air temperatures close to 20°C and 10% of the normal rainfall. Warm settled conditions favour algal blooms.

When moving along the canal, it was apparent that some of the blooms had crashed and the water had turned a blue colour, sometimes with a milky white appearance. It was also noticed that the apparently toxic reaches of the canal were amongst those recently dredged by British Waterways some months previously, although no fish mortalities had occurred during their operations. The fact that there had been substantial rainfall on the night of the 3rd/4th March 1998 may have led directly to the bloom crash. Or alternatively, something might have been washed into the system to then cause the bloom to crash. Canal dredging sediments were taken for chemical analysis, but no chemicals emerged as being of concern.

3.2   Ecotoxicological Assessments of the Kennet and Avon Canal

3.2.1   State of toxicity in the system   Although toxicity had been clearly seen in the canal system, resulting in the deaths of over 30,000 fish there, current toxicity levels, some 8 days after the event, were not known. The canal had remained closed so it was likely that toxicity remained. Deployments of roach (*Rutilus rutilus*) and chub (*Leuciscus cephalus*) (cyprinid fish similar to those fish killed in the canal) were made in keepnets into several of the canal reaches. Fish were examined twice daily for signs of erratic behaviour, morbidity, and death. Fish were also deployed in "control" reaches upstream of the toxic events. Early evidence came within 2 days when fish began to show signs of "sluggish" behaviour and in some cases mortality. Some of the canal reaches were still toxic. Roach were shown to be less sensitive than trout or chub to the toxic water. Deployments continued over the next three weeks, so that toxicity could be monitored over the time.

In parallel to the field deployments, large water samples were taken back to the laboratory to run fish exposures in the laboratory. The field deployments monitored the environmental toxicity *in situ*, but this can be affected by environmental factors such as temperature, physicochemical parameters (pH, dissolved oxygen) water circulation or daylight. Laboratory 96 hour exposures on trout and chub would assess the time to event (behaviour, morbidity, death) under more managed conditions and with clean water controls, although the field sampled water used would only relate to a single sampling occasion.

Similar results were found in the laboratory to those in the field. Strikingly, high mortality levels occurred within 17 hours at some sites although at some sites toxicity was lower . Clearly, toxicity to fish remained high. Other toxicity evaluations (Water - *Daphnia magna* 96h lethality test; Microtox®; Sediment – fish, *Chironomus riparius*) confirmed the field observations that invertebrates were not sensitive to the toxicants and that sediments were not apparently directly toxic to either invertebrates or fish.

What became a highly significant finding was discovered following the sampling of water for laboratory exposures on 17 March. In the first laboratory exposures, samples from one site had showed no toxicity and none had been previously seen in the field, making it effectively a control site. In this second experiment some 5 days later, after 17 hours after exposing the fish, over 90% had died. The water samples had been left overnight prior to starting exposures the next day, during which time the algae and other solids had settled out of suspension. In had become apparent that this was a dynamic event – toxicity was emerging before our eyes. In fact, the field event was repeating itself in the laboratory. Our attention focused on possible biological causes.

Field deployments continued for 3 weeks. Toxicity fluctuated in the canal reaches during that period. With toxicity evident, the canal remained closed and the dynamics of the biological changes became more clear. Canal reaches with algal blooms were non-toxic or significantly less toxic than others. Canal reaches became toxic as the bloom was crashing, the water turning a milky (sometime blue) colour. The colour was thought to be related to *Pseudomonads* (phycocyanin pigments) and the white calcium carbonate precipitations. These observations indicated that the algae themselves might be toxic when breaking down, or that bacteria feeding on the algal products were themselves toxic, or produced a toxin in response to the food source

3.2.2 Development of remedial measures   Having some clue as to the cause opens up a number of avenues for further work. The key priority was to remediate the canal system and allow it to be reopened. Using toxic canal water and the 96 hour fish bioassay, a number of sample manipulations were undertaken to the toxic water - filtration; autoclaving; addition of bentonite, powdered activated carbon, iron flocculant, hydrogen peroxide.

No mortality was observed in any of the treatment groups, whereas 100% mortality was found in the unmodified water after only 24 hours. The fact that filtration removed toxicity indicated that the toxin was either bound to particulates or a large molecule which itself binds to the filter matrix.

In taking these findings to the field situation, hydrogen peroxide dosed at 5mg/l was considered to be the most effective option that could be applied.

3.2.3 Application of remedial measures Hydrogen peroxide is sometimes used in remediation of fish mortalities where the water body has a low dissolved oxygen concentration.   Certain pollution events can lead to substantial reduction in dissolved oxygen (e.g. silage, milk or other foodstuffs) and hydrogen peroxide does provide instant dissolved oxygen, whilst also breaking down organic particles because of its strong oxidising nature.   Although clearly toxic to life, there are no reported harmful effects to fish when mixed into water in this way. However, it is not easy to use and requires health and safety precautions for operatives in its delivery into the water.

In order to treat a waterbody effectively, the liquid hydrogen peroxide needs to be well-mixed in with the water. Treating several kilometres of canal is therefore a challenge. Drums of hydrogen peroxide were placed onto boats with outboard motors. The liquid was poured continually into the canal; whilst the boat continually crossed over the canal water from bank to bank, the motor propeller itself aiding mixing. Fluorescein dye was mixed in with the water to provide visual indication of effective mixing. Once the treatment had been completed, fish were again deployed into keepnets in these reaches and toxicity assessed for a further 96 hours. As no toxicity was found in the canal – for the first time in 4 weeks, the Kennet and Avon canal was reopened on 3 April 1998, one month after the incident was discovered.

3.3   Further Investigations into the causes

3.3.1   Toxicity Investigation and Evaluation   As the event was progressing towards the conclusion that it was a biological cause, it was becoming increasingly clear that that we knew very little about the toxin. It would be important also to characterise the toxin. Moreover, it would be prudent to gather more supporting information about the biological cause whilst samples were still available.

A Toxicity Identification and Evaluation (TIE) exercise was commissioned to assess how the toxicity varied following various manipulations to the water sample. TIE is a stepwise process whereby a sample is treated to a number of chemical (e.g. extractions; treat with agents with specific chemical properties) and physical (heating/cooling/filter) treatments in order to change or apportion toxicity to a particular characteristic of the substance. Further manipulations might resolve the substance into a particular fraction for chemical analysis, for example, by chromatography. At each manipulation stage, the sample is re-tested against a selected biological test, in this case, a simple fish toxicity test, to see whether toxicity has changed in any way.

The manipulations indicated that virtually all treatments affected toxicity. Heating, freezing oxidising agents and adjusting pH all rendered the sample non-toxic, indicating little chemical stability of the toxin. Loss of toxicity at as low a temperature as $40^{\circ}C$

would be a strong indication a toxin of biological origins. Various extractions followed by re-suspension also led to the loss of toxicity, indicating that the substance was not a particle-bound micropollutant, and effectively is not a robust molecule. Only centrifugation, aeration and EDTA chelation treatments failed to remove toxicity, indicating that the toxin was not particle-bound; non-volatile and not a cationic metal ion.

Although we know something about what the toxin is not, its identify still remains elusive. Its inability to be manipulated in any way defied chemical analysis, including specific analysis that was targeted including lipopolysaccharides.

3.3.2 Fish histopathology The diagnosis of gill damage as the primary site of action of the toxin was an early observation. Part of the fish trial evaluations was to examine gill tissue as one of the endpoints of the experiments, to confirm that the canal was still capable of causing toxic effects (if mortality had not occurred within the 96hours). Scanning electron microscopy proved to be a useful tool to enable examination of the hyperplasia condition. Not all fish that in the experiments died but most were showing the characteristic signs of damage.

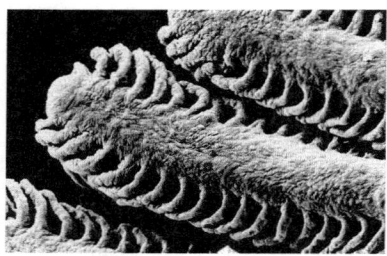 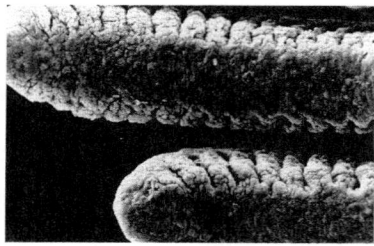

Figure 4   *Scanning electron micrographs of fish gills (left – normal; right showing gill hyperplasia)*

This characteristic form of damage remains one of the key diagnostic endpoints for this toxic phenomenon.

3.3.3 Microbiological assessments Microbial agents, particular bacteria, could have contributed to the fish deaths in a number of ways, by
- Direct infection by pathogenic bacteria present in the water and sediment
- Infection of stressed fish by opportunistic pathogens in the water or sediments
- Release of endo- or exotoxins by bacteria

There is a literature on some previous fish mortalities caused by *Aeromonas, Pseudomonas* and *Vibrio*. However, this mortality was not an apparent infectious agent. Indeed the pathology indicated general prior good health of the fish that died. The evidence points to an endo- or exo toxin, released by another group of bacteria. Endotoxins are usually associated with the lipopolysaccharide (LPS) complex associated with the outer envelope of Gram negative bacteria such as *E.coli*, *Salmonella Shigella and Pseudomonas*.LPSs can elicit a number of inflammatory responses in animals. Although some endotoxin is released from the cell wall, for the most part endotoxins remain associated with the cell wall until disintegration of the bacteria. Endotoxins tend to be less potent than exotoxins but are heat stable. Since the TIE study indicated poor thermal stability, it is unlikely that an endotoxin was involved in the mortality.

Exotoxins are often very potent, retaining high activity at very high dilutions. They can resemble enzymes in a number of ways in that being proteins they are denatured by heat, acid and proteolytic enzymes, have a high biological activity and exhibit specificity of action. Some toxins have very specific cytotoxic activity, whereas others have broad toxic action, for example cleaving phospholipids leading to cell lysis. Of the exotoxin-producing bacteria likely to be present in watercourses, *actinomycetes* contain a number of exotoxin-producing genera, such as *Corynebacterium*, which do not have to be a dominant member of the microbial community to have a toxic effect. They are also associated with sediment Of all the possible causes; a microbial exotoxin was hypothesised to be the most likely explanation for the initial fish kill at Berkshire Trout Farm.

4 CONCLUSIONS AND FURTHER INVESTIGATIONS

4.1 Overall assessment of the Hungerford Fish Mortality

The event was believed to be the first of its kind. Still, warm conditions led to the development of unusual algal blooms for the time of year. A rainfall event caused the bloom to crash and then followed by the increase in bacterial flora to produce sufficient quantities of a bacterial exotoxin to kill over 500,000 fish.

But it was noted that toxicity was primarily occurring in the dredged canal reaches. How significant was this factor in the event? It is likely that dredging increased nutrients such that the favourable conditions existed to support the diatom bloom. Dredged material was placed on the river banks in the autumn, with the consequence of it falling back into the canal when it rained. Conspiring with the warm weather, the dredgings would have active microbial communities, likely to be at higher numbers than in the colder water.

The explanation for the event therefore needs to encompass the role of dredging activity. It is possible that dredging, as well as increasing the trophic state of the canal, led to the toxicity in the canal following the rainfall, by washing directly back into the canal and either

- containing a large enough pulse of toxin to cause the fish mortality, and/or
- that the microbial community in the dredgings formed an inoculum for continued growth in the canal, leading to the production of greater amounts of toxin in the canal over the days and weeks later.

This is to some extent speculation. But the hypothesis is testable and research is now underway to provide the necessary data. A key question remained – is the event going to recur, or was it a one-off?

4.2 Lessons learned

This "pollution" event was a great learning point for the scientists involved. Key lessons were:

- look for obvious causes based on experiences and likely risks in the catchments
- keep an open mind to biological as well as chemical causes of mortalities
- separate out the roles of technical "investigation" from the "incident management", and have dedicated resources focused on managing the flow of technical information (particularly with so many technical support staff involved).
- Not to rely solely on analytical chemistry to solve the mystery
- Use bioassays to identify locations of toxicity and its severity, not analytical chemistry
- Biological testing and TIE delivers the key answers – its time and cost effective

- Take large quantities of samples in the early stages of an event – both chemists and biologists need sufficient material to work with.
- Seek specialist advice and support early, particularly as this event was not straightforward.
- Liase early with those in the locality affected by the event and other relevant stakeholders; and think about the information they need – you may not be undertaking work to provide it at that time.
- Have the confidence of your convictions to implement the remedial solution, however big the scale of the task is.

4.3 Post-Script – what happened next?

The 1998 event will live long in the memory of those who live in the Hungerford area, those who were affected and worked to solve the incident. Nobody wanted a repeat incident. In 1999, remote monitoring stations were set up on the canal system to assess a number of important chemical and biological parameters including chlorophyll a, pH and dissolved oxygen as a means of watching for the signs of algal blooms occurring. Fish monitoring was also continued to asses for signs of stress. In both 1999 and 2000, algal and fish activity indicated the potential for another fish mortality and the canal was dosed again with hydrogen peroxide to remove the toxicity. Bacterial samples were collated and several *actinomycete* species have been isolated as being present at these events.

As awareness of the event and its circumstances grew in the subsequent years, the Agency developed a protocol for sample collection for field staff concerned that they may also experience similar events in their locality. This has had remarkable results. Since the initial event in 1998, Agency staff have now attended more than 30 suspected impacts, including some large fish mortalities in still waters. Mortality events are running at around 10 per year. Fish tissue samples have been archived as part of a continuing research investigation, used initially to characterise each event as being caused by fish gill hyperplasia. Microbiological samples are also being taken for identification as for innocula for laboratory fish exposure experiments. Recent research has successfully induced toxicity to fish in laboratory conditions from bacterial isolates. This will be published in the literature over the next few years.

Much work remains to be done, but it would seem that the from the initial fish mortality at Hungerford, a wider phenomenon has been discovered and investigated. Mystery fish kills have often been put down to unknown pesticides, physical damage of the fish, low dissolved oxygen and sudden pH changes. Perhaps these events did have a logical, man-made cause. But this case study has shown that we ignore biological agents as environmental toxicants at our peril.

Acknowledgements

This investigation was a combined team effort between a number of scientific groups within the UK.

Environment Agency: Science Group; West Area, Thames Region; National Laboratory Service
Water Research Centre, Medmenham
University of Lancaster
University of London, Royal Holloway College

David Bucke Pathology Services
Freshwater Biology Association
University of Dundee
Centre for Environment, Fisheries, Aquaculture Sciences

References

1   Environment Agency (1998) Technical Investigation of the Hungerford Fish Mortality. *Technical Report* CO4626 (Available from the WRc Bookshop **44 (0)1491 636500)

2   Environment Agency (1998) Technical Investigation of the Hungerford Fish Mortality. Annex to Technical Report. *Technical Report* CO4627 (Available from the WRc Bookshop **44 (0)1491 636500)

THE FOOT AND MOUTH DISEASE OUTBREAK 2001

G. Bateman
Environment Agency

1 INTRODUCTION

'The Foot and Mouth Disease (FMD) epidemic of 2001 was one of the largest in our history.........It is worth making two key points at the outset. First given the wide spread of the disease throughout the country prior to detection, the impact of this outbreak was bound to be very severe. Even had everything been done perfectly by all those concerned to tackle the disease, the country would have had a major epidemic with massive consequences. Second, many farmers, local people and government officials made heroic efforts to fight the disease and limit its effects. Through their efforts it was finally overcome and eradicated after 221 days, one day less than the epidemic of 1967-78'.[1]

The Department of Environment Food and Rural Affairs (DEFRA), previously the Ministry of Agriculture Fisheries and Food (MAFF), is responsible for dealing with animal health matters, including an outbreak of foot and mouth disease (FMD), which is a notifiable disease of animals. DEFRA is also responsible for the sponsorship of the agriculture and food industries. This department therefore took the lead on FMD policy and operations and worked closely with the National Assembly of Wales (NAW) and the Environment Agency to solve disposal difficulties and incorporate environmental advice into developing disposal policy.

The Environment Agency has a duty to prevent and control pollution and to report on the state of pollution of the environment in England and Wales. Throughout the outbreak we worked with others to minimise the risk of environmental harm and in the context of this conference I will concentrate on the impacts on water pollution and the lessons we learned.

2 THE OUTBREAK

On 19 February 2001, a veterinary inspector from the State Veterinary Service of the Ministry of Agriculture, Fisheries and Food while carrying out a routine inspection of an abattoir in Little Warley, near Brentwood, Essex noticed vesicular lesions on 27 sows and one boar. The following day analysis identified FMD in the animals. The virus was of type 'O' subtype Pan-Asian, which was identical to virus in recent outbreaks in South Africa. The abattoir was a major meat exporter, sourcing animals from more than 600 farms throughout the UK, including Northern Ireland.

Movement restriction orders were placed around the abattoir and also around farms in Stroud in Gloucestershire, Great Horwood in Buckinghamshire and Freshwater Bay on the

Isle of Wight. The most likely source of the disease was quickly pinpointed by MAFF inspectors to Burnside Farm at Heddon-on-the-Wall, Northumberland. An inspection revealed that the disease had been present in stock "for about 14 days". Forty sheep from a neighbouring farm had been taken to Hexham market on 13 February, when the disease would have been present on Burnside Farm. There are indications that these sheep were infected from that farm. More than 3,500 animals were sold at Hexham market and distributed throughout the UK. The original 40 sheep went to a farm in Devon.

By 23 February six outbreaks had been confirmed in Essex and in the North East. The Agriculture Minister, Nick Brown, announced a nation-wide ban on the movement of all livestock except under licence; exports had already been banned on 21 February. That weekend, MAFF commenced disposal of animal carcasses by incineration on pyres built from straw, coal and railways sleepers, using methods employed during the 1967 outbreak. Animals were slaughtered on site and carcasses burned on pyres built near to the point of slaughter.

There were 400 outbreaks within 4 weeks and 2030 outbreaks in total. But this was not the whole picture. For each infected premise (IP) there were several contiguous premises which were culled to deliver a 'fire break' to stop spread of the disease. The role of the Environment Agency was to assist in safe disposal of the bodies, including those carcasses from the livestock welfare disposal scheme (LWDS) that were a consequence of the welfare impacts of the animal movement ban. In total some 4.5M animals were culled as a result of disease and some 2M animals from the LWDS.

The main areas affected were Western and Northern England, Wales, and Southern Scotland, which fortunately were not the most sensitive groundwater aquifers used for public water supply.

To set the context, the UK livestock herd before the outbreak comprised 11M cattle, 8M pigs, 43M sheep and the FMD cull was in the same order as the annual natural mortality, including BSE control. But this was unevenly spread across the country and the bulk of disposal was within in a two month period.

3 REGULATORY BACKGROUND

The primary pieces of legislation enforced by the Agency during the outbreak were the Environment Protection Act 1990, Water Resources Act 1991 and Environment Act 1995. The Groundwater Regulations 1998 permit control of discharges of listed substances to ground and require prior investigations and requisite surveillance. The Animal By-Products Regulations were enforced by MAFF and incorporate guidance on animal disposals encompassing the code of good agricultural practice for protection of water. This latter guidance had hitherto been applied to disposals of carcasses and fallen stock not greater than 8 tonnes. There was also a debatable need for planning permission but this was waived in the context of a national emergency.

4 INCIDENT MANAGEMENT

The Agency played a major role in the strategic and local management of the outbreak by raising environmental issues at Cabinet Office Briefings (COBR) and at the Joint Co-ordination Centre (JCC). There were close links between the National Assembly for Wales and the Agency, including the full time presence of Agency staff at the National Assembly Incident Room. A significant level of staff resource was involved at National, Regional and Area level in incident management. A National Foot & Mouth Task Group managed national co-ordination of the incident.

The Environment Agency implemented its National Incident Management procedure and opened the National Incident Room on 22 February 2001. Our incident management procedure includes a National Base Controller, Regional Base Controller and Area Base Controller. These roles were established to provide links between MAFF at Page Street and then subsequently the Cabinet Office Briefings, the Joint Co-ordination Centre and national government departments, regional organisations and local councils and site-specific issues respectively.

The national incident management command and control structure as it related to the Agency's involvement is shown in Figure 1.

Figure 1   National incident management command and control structure

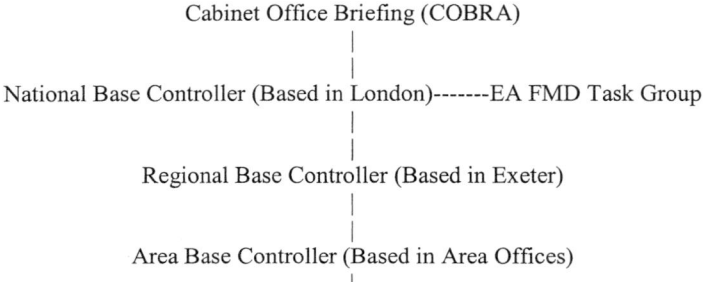

Cabinet Office Briefing (COBRA)
|
|
National Base Controller (Based in London)-------EA FMD Task Group
|
|
Regional Base Controller (Based in Exeter)
|
|
Area Base Controller (Based in Area Offices)
|
Links with local MAFF (DEFRA), DETR DoH Army, Local Authority Environmental Health Officers, English Nature, NFU, Water Services plc's)

We provided MAFF (DEFRA) with assistance in the field, administrative support at MAFF (DEFRA) offices, provision of technical advice on environmental matters and attendance at council and public meetings. We also attended regular local and national briefings with key organisations to ensure continued communications including MAFF, Waste Management Operators, MoD, Environmental Health, NFU and English Nature and gave several media interviews and press conferences.

At the height of the outbreak in late March/early April, over 40 new cases were confirmed per day. During this period an estimated 400 Agency staff were engaged in work associated with the outbreak, responding seven days a week. The number of confirmed cases by Agency Area and Region are available.

As the number of cases of FMD increased, the pressure to respond also increased. At the same time public interest and opposition to disposal, via a number of routes, also increased putting an increased burden on Agency resources. In addition, there were a number of water related pollution incidents associated with FMD, mostly associated with disinfectant and leachate. The environmental impact of the outbreak and measures to deal with it are reported in "The Environmental Impact of the Foot and Mouth Disease Outbreak – Preliminary Assessment – November 2001 *(*www.environment-agency.gov.uk*).*

5 DISPOSAL OF CARCASSES

There were a number of critical issues for disposal. The vets recommended that carcasses should be left on the surface for the least possible time, in order to reduce risks to health and the environment. The vets also required infected animals to be killed within 24hours and those at risk within 48hours to prevent disease transmission. This meant that the volume of carcasses requiring disposal quickly outstripped the traditional disposal routes. Rapid removal from farms was required, but the rapid escalation of epidemic was set against the logistical problems of time to assess environmental risks and construct pyres, burial pits etc. Also to arrange transport to disposal sites, the resources were huge.

The Environment Agency and MAFF/NAW agreed joint working arrangements and principles for disposal activities. The disposal of carcasses was subject to the Animal Wastes Directive and the Animal By-Products Order 1999. A hierarchy of preferences for disposal routes for carcasses, was developed:

• rendering at an appropriately authorised and monitored rendering plant (highly effective in destroying the virus; existing plant was well equipped to deal with large numbers of carcasses)

• incineration in authorised and regulated incinerators (highly effective in destroying the virus and carcass; can deal with large numbers of carcasses, but few incinerators available)

• landfilling in appropriately engineered and authorised landfill sites (engineered containment minimises risks to groundwater and the wider environment, and appropriate sites are suitable to accept relatively large quantities of carcasses)

• burning on the farm (low risk of virus transmission as carcasses are not moved from the site, but may give risk to short term air quality and odour issues; supervision required to ensure combustion is complete; construction of pyres is time-consuming)

• burial on the farm (quick and simple to undertake, but significant potential risks to groundwater from leachate from burial pits, and potential for land to be blighted)

Disposal was largely dictated by the availability of local options and logistical support, strategies varied across the UK.

The Agency was fully aware that MAFF/NAW would have to take a number of issues into consideration when deciding the optimal route for each case of disposal of carcasses during the outbreak. The agreed disposal hierarchy was not rigid. The Agency applied proven risk assessment techniques combined with pragmatic, local decision-making in close co-operation with MAFF and logistics teams from the Ministry of Defence (MoD).

6 ADVICE ON DISPOSAL OPTIONS

The Environment Agency was represented at the national level of strategic planning for disposal, and at local level in all affected parts of the country. Agency staff liaison officers were based in local MAFF offices to act as key contacts between those organisations involved. In addition the Environment Agency co-ordinated the concerns of the Water Industry, English Nature and other environmental groups and consulted locally within short timescales on environmental risks. A great deal of advice was provided internally and externally through our intranet and internet sites.

All advice on local disposal options was handled by local-based Agency teams, where necessary working seven days a week, using their expert knowledge of hydrogeological and environmental factors. Agency teams aimed to recommend safe disposal routes within three hours upon faxed consultations from MAFF/Army. This target was achieved, with the exception of a small number of complex cases. Under the UK Groundwater Regulations the Agency developed a fast-track authorisation process and issued some 3000 authorisations to MAFF/DEFRA, mainly for the disposal of cleaning and disinfection materials. As a result DEFRA continues to undertake surveys and requisite monitoring of sites to ensure pollution is not occurring.

There were many more sites where the Agency was unable to authorise burial on farm because of the direct risk of pollution of public and/or private water supplies. For example in Devon burial of carcasses was rarely permitted due to the high water tables following a record breaking wet winter.

7 PROTECTING THE ENVIRONMENT

Amongst the Environment Agency's main concerns were the potential environmental and public health risks of groundwater pollution. These were not issues in all areas, but were a significant risk where aquifers[1] are used for abstraction of water for public water supply or food/drink processing, or where they support private boreholes and wells used for drinking water.

Pollution risks from carcass burial, and to a lesser extent from burial of ash following burning, relate to potential leaching into groundwater of the breakdown products of carcass decomposition, rather than any risk of the transmission of the Foot and Mouth virus. Potential pollutants are ammonia, chlorides, phosphates, degradable organic compounds (eg fatty acids) and bacteriological contamination. Metal concentrations could be elevated and there could be an impact on the taste and smell of water.

In England and Wales, approximately 35% of public water supply is based on abstraction of groundwater. The Agency's maps of *Groundwater Vulnerability in England and Wales* show the location of aquifers in England and Wales.

In Cumbria, for example, where the largest number of confirmed FMD cases occurred there are some 140 abstractions for public water supply and/or food or drink processing. In addition, several thousand private water supplies rely on groundwater. Contamination of groundwater can be difficult (and sometimes impossible) to remedy. Secure water supplies are important to the recovery and long term viability of the rural economy.

8 POLLUTION INCIDENTS

During the foot and mouth outbreak approximately 200 pollution incidents were recorded by the Environment Agency, as a direct result of foot and mouth activities. The majority of pollution incidents related to blood and body fluids from either slaughtered animals left in fields or from vehicles carrying slaughtered animals from farms to either landfill or rendering facilities. A number of reports were related to concerns regarding the pyres that had been used to burn carcasses. In addition to pollution incidents there were a huge number of complaints about odour from certain landfill and mass burial sites.

There were four category 1, major pollution incidents. The first in the South West, where an overflowing slurry tank on a farm within an infected area resulted in the death of over 350 fish. The farmer blamed the inability to spread slurry as a result of disease restrictions. The second in Wales, where a significant leak of disinfectant from an abattoir on Anglesey resulted in thousands of eels and fish dying. Thirdly, in the Midlands slurry and disinfectant wash waters were discharged to a watercourse causing a major fish kill. The Environment Agency is considering legal action in these cases. The fourth incident was a discharge of slurry from a farm near Carmarthen where many fish were killed. Enforcement action has been taken.

---

[1] *Aquifer: a porous layer of rock, which soaks up water as it percolates down through the soil after rain, and acts like a natural underground reservoir. Aquifers are usually at considerable depths. The water is usually pure due to the filtering effect of upper layers of soil and rock. Where aquifers of suitable quality exist, it is usually practical to drill boreholes to give access to the water.*

The environmental risks posed by this outbreak as opposed to the one in 1967 were different and much greater both in scope and scale. Much more is now understood about groundwater and contamination and a regulatory regime exists in the form of the groundwater regulation, which is aimed at preventing pollution of groundwater. The disease was countrywide and the scale was greater posing greater pressures on disposal routes for carcasses, pyre ash, disinfectant and associated disease risk materials.

There was BSE in the cattle herd and the Environment Agency advised its staff not to allow burial of any cattle. This advice was later modified following advice from the Spongiform Encephalopathy Advisory Committee (SEAC) to older cattle i.e. those born before 1 August 1996.

The main disposal route for older cattle was rendering or burning on pyres. This resulted in much work to assess the BSE risk from pyre ash. However, analysis demonstrated that the risk from BSE in the ash was much diminished.

9 MONITORING ENVIRONMENTAL IMPACTS OF FMD

One of the Agency's most important roles is monitoring and reporting on the state of the environment in England and Wales. Environmental monitoring during the outbreak was severely curtailed. Our current scientific evidence is therefore minimal. However the fact was that many of our staff found themselves trapped within infected areas at the start of the outbreak. These staff were classified as 'dirty' and were working with MAFF vets, contractors and the army in the infected areas and on farms, so could directly influence disposal and disinfecting practices to ensure pollution control measures were undertaken.

Where pyres were used to burn carcasses, smoke contributed to poor air quality in the short term. Whilst local air quality is essentially a matter of local authority control, the Environment and AEA Technology undertook monitoring at several pyre sites on behalf of MAFF/DEFRA and the Department of Health. Results are available on the Agency website and these were shared at the time with relevant local authorities and interested bodies where pyres were used.

10 AGENCY ADVICE

In addition to dealing with carcass disposal issues, the Agency gave advice and assistance to MAFF/DEFRA, the farming community and others on:
- the use and disposal of disinfectants
- accelerants for burning
- disposal of slurry, which is normally spread on land
- disposal of milk which cannot be sent to normal market or distribution systems
- disposal of blood/guts contents from abattoirs
- disposal of sewage sludge
- access to the countryside for environmental works

The Agency's own routine environmental and flood defence activities were affected by the outbreak, with some work stopped in areas where access would pose a risk of transmission of the disease. The main impact was that Agency staff could not attend some pollution incidents, and there were delays to some flood defence works. Guidance on assessing the risks of carrying out activities was issued to all Agency staff to enable work to go ahead as normally as the situation allowed.

The Agency advised anglers to take a responsible approach when deciding where to fish, in line with Government advice. Similarly, the Agency and the British Waterways Board issued advice to boaters and recreational waterways users.

11 AGENCY STAFF

The Agency seconded a considerable number of staff to MAFF to assist with the heavy workload brought about by the outbreak. Their roles varied from administration and data handling staff to acting on site as temporary animal health officers.

12 LESSONS LEARNED

a) Staff are the most important resource. They need to be cared for, properly developed and trained. They are required to be flexible, committed and must be allowed to make their own decisions on the ground in the context of broad policy frameworks.

b) Regulatory powers and influence need to be focussed to ensure protection of the environment. To date, in spite of the scale of the outbreak, the disposal of carcasses and ash, the use and disposal of disinfectants and disposal of associated wastes has not caused serious environmental problems. So we were successful in using risk assessment methods to minimise the impact.

c) The initial response to any emergency should be rapid and broad. The integrated response to the outbreak by MAFF in the first month was slow. MAFF vets were in emergency mode from 20 February, but were overwhelmed by the need to carry out a multiplicity of tasks to cope with the scale of the outbreak. A disease outbreak on this scale had not been taken account of in scenario planning or contingency plans. The Environment Agency was fully operational in emergency mode from 23 February 2001, but it was not possible to arrange a meeting with MAFF Head Office until 6 March to discuss the disposal strategy and environmental concerns. However, there were early meetings with the water industry to discuss the impact of the outbreak on access to farmland for water treatment, sewage sludge disposal and routine operations.

d) Communications is crucial. Communication equipment must be adequate and robust. This applies to telephones, computers and statistical tools/models. Communication between Agency officers and MAFF officers was very slow in the first few weeks. It was not easy to contact key MAFF personnel at regional and national level to discuss the emergency response, or answer inquiries, for example, telephones were permanently engaged and faxes and e mails were not answered in the first few weeks. But it was clear that there were huge individual efforts and hard work by everyone involved. Communication improved when the army arrived and additional resources were deployed.

e) The communication of policy and the implementation of action depends on clear command and control. This has now been addressed in the draft contingency plan for dealing with an outbreak of FMD which can be found on the DEFRA website. (www.defra.gov.uk). The deployment of Regional Operations Directors (RODS) helped co-ordinate action on the ground and the quality of information collected to inform policy makers.

f) Risk-based techniques. Animal diseases need to be managed in the context of the wider impacts by applying risk-based techniques to policy development, environmental monitoring and operational responses. The protection of people and property from flooding, and environmental and public health protection and monitoring should be subject to adequate risk-assessment to ensure appropriate decisions are taken during an animal disease outbreak.

g) The full understanding of roles and responsibilities of partner organisations during an emergency event is fundamental. This was not the case at the start of FMD epidemic.

h) Information needs to be 'real time' for effective planning and policy achievement. As the outbreak escalated the systems could not keep pace and in particular the ability to extract data from the field was inconsistent. Busy people dealing with the operational pressures of the outbreak were unable to provide accurate data from the ground. This was demonstrated by the Devon experience where staff were reporting a backlog of carcasses on the ground which was giving cause for concern. Operational decisions were made in respect of mass burial and disposal to deal with the backlog, when in reality the problem was the collection of statistics and the local staff were already disposing of carcasses. This resulted in the building of Ash Moor mass burial site, which was not subsequently used.

i) Environmental information is important. The details of locations, content and methods of disposal need to be collected at the time to ensure strategies to protect public health and the environment are appropriate. Such data has been hard to collect retrospectively to determine the actual or potential longer term risks. The Agency recommend, for future animal disease outbreaks, the development of information and data sharing strategies that are supported from the ground upwards, adequately resourced, accurate and timely. there is a need for a robust information system that is accessible by all relevant partners.

j) Considering those affected. Many Agency staff on the ground acted in support of MAFF vets and contractors. They also spent time with farmers, their friends and neighbours and supported them following the culling on farms. This contact was desperately needed and was absent from any contingency plans. Counselling and support is needed for victims and helpers in any future contingency planning.

k) Disposing of the problem. The lack of disposal options lead to the development of mass burial sites in the infected areas. The Agency assessed some 400 sites put forward by the army and MAFF and recommended a handful for further investigation, which were subsequently used. Environmental risks remain with the future management of mass burial sites and the treatment and disposal of leachate. These risks need to be minimised. The agency is working with MAFF to ensure that the standards applied to the mass burial sites at Great Orton (Watchtree); Tow Law; Widdrington; Throckmorton; and Ash Moor (currently empty), are similar to that required for licensed landfill sites. The Agency recommend that, so far as practicable, existing mass burial sites should comply with pollution prevention control standards and a clear statement is made regarding who is responsible for their long term management.

l) Future outbreaks will happen. In the draft DEFRA Contingency Plan the disposal options have changed, to include incineration as first option followed by rendering and use of licensed landfill. This reognises the need to use properly licensed facilities to their maximum potential before considering other routes.

m) Waste minimisation should be considered in the contingency planning stage. Culling on farm is recognised as the most effective disease control measure, but the slaughter and disposal should be co-ordinated to ensure efficient removal of carcasses. Any methods to reduce the volume of production of carcasses, whilst achieving elimination of disease, should be considered further. Vaccination is an option that would achieve the objective from an environmental standpoint, by keeping stock standing to allow subsequent slaughter through licensed disposal routes. Vaccination is a complex issue, but use of disease free or vaccinated animals for food is clearly preferable to disposal as waste. The sustainable options are to minimise the production of waste and to seek economic solutions to reduce, reuse and recycle and only dispose of carcasses in the last resort. Measures to manage animal disease outbreaks in future should consider matching disposal volumes and disposal capacity.

n) <u>Policy development at a senior level is needed</u>. The Cabinet Office Briefing Room (COBRA) was the place where policy was developed and decisions taken, particularly when the army or ministers were present. it served as an invaluable co-ordination of government departments. It was sometimes hampered by the lack of timely and accurate information from the ground.

o) <u>Policy decisions must be informed and implemented</u>. The Joint Co-ordination Centre (JCC) at Page Street was set up by MAFF and the army and jointly chaired. The principles of operation were: -

to create and maintain an accurate ground picture
to create an all informed 'network'
to facilitate passage of information flows, information management
and dissemination of instructions.

The JCC provided a very efficient and effective means of sharing information, raising concerns and directing field action. it provided a positive impact and the capability to focus on priority issues. the 'battle-rhythm' of three 'bird tables' a day injected purpose and pace into dealing with the disease.

p) Research and development. More needs to be known about the behaviour and spread in the environment of the FMD. There were several models for predicting the infective behaviour of the disease, but little was available in relation to disposal strategies and the long term impacts of organic pollution of groundwaters. Not enough is known about the fate of leachate and its impacts on groundwater. More research is needed to model groundwater flows, air pollution dispersion models and the risks associated with BSE in the natural environment. More research is needed into the impacts and pathways of pollutants such as disinfectants. Excellent work was done by environment agency staff in our National Centres and Services and on the ground. Also many contractors such as BGS and AEA provided vital support in assessing environmental risks, but more work is needed

13 CONCLUSION

I believe we are better prepared, have learned the lessons better and better understand the environmental consequences of animal disease outbreaks. But all emergencies depend on the state of preparedness the degree of planning and the quality of those involved from top to bottom. We must ensure we continue to resource the plans, exercise them and be prepared.

For further information about the Environment Agency's involvement in dealing with the Foot and Mouth outbreak can be found on our website at <u>www.environment-agency.gov.uk</u>

Other useful sources of information:

For information about Foot and Mouth disease and measures to control it:
<u>www.defra.gov.uk</u>
<u>www.doh.gov.uk</u>
<u>www.fsa.gov.uk</u>

Disclaimer.
The views expressed in this presentation are those of the author and not necessarily those of the Environment Agency.

References
1   I. Anderson, Foot and Mouth Disease 2001:Lessons to be Learned Inquiry.

BURNCROOKS DIESEL INCIDENT

JOHN FAWELL

1 INTRODUCTION

In December 1998 West of Scotland Water were a relatively new organisation having been removed from direct local authority control. Burncrooks is a surface water abstraction and drinking water treatment works high in the catchment and an important supply providing about 23.5 megalitres per day.

2 THE INCIDENT

On the 9th December 1998 Scottish Hydro brought a temporary generator on-site at West of Scotland's Burncrooks WTW to allow work to be carried out on the main supply in the area. The arrangement was made locally and WoSW management were unaware. In addition the senior operator at the site was sick and was not at work. The generator was sited in a bunded area in the centre of the works close to the buildings.

The electricity company employee left the generator immediately after start-up to make a phone call. In his absence there was a malfunction and diesel was pumped out through the air breather. When the electricity company operative returned he corrected the problem but by then an estimated 10 to 80 litres of diesel had been spilt. The incident was reported to Scottish Hydro and SEPA because of concern that drains led to the nearby burn. By this time it was dark and the weather was deteriorating. The local WoSW operative was informed, but WoSW management were not. There was no sign of diesel in the burn but it was dark and very wet and the burn was fast flowing.

Taste complaints were received the next morning at the same time, as there were high chlorine residuals at the works. The site operative reported exhaust fumes in the works but did not confirm that the smell was indeed due to exhaust fumes from the generator. As the number of complaints increased management declared an emergency declared although they had no clear cause for the problem. An investigation at the works found that surface water from the bunded area drained to the wash water tank that was recycled to the works.

As a consequence diesel had penetrated the filter media and the low molecular weight, relatively water-soluble aromatic molecules that are present in diesel were being pumped into supply. A number of these aromatic hydrocarbons have a very low odour threshold and can render drinking water extremely unpleasant at concentrations of less than 30 μg/l. The result was that highly odorous water was being supplied to a large area with about 27,000 domestic properties and 66,000 people. This was supplied by 341 km water mains. A faulty valve interfered with re-zoning and widened the affected area.

3 ACTIONS TAKEN

Consumers were told only to use the water for toilet flushing and bowsers with alternative supplies were provided. The process of informing such a large number of consumers was complex and did not go smoothly resulting in some confusion and an increase in complaints. It was also a bad start for media involvement and the media, who had not been brought on board sufficiently early, took a very hostile stance. The Chief Executive was on his first holiday in a significant period and was unable to return immediately due to the lack of availability of flights and this was used by the media to further inflame the situation. Christmas was rapidly approaching and there was concern that the incident would not be resolved in time.

The source of the contamination was actually in supply and could only be removed by replacing the contaminated filter media in all of the filters and cleaning the system. This could not be achieved immediately and so there was a requirement to manage the incident over an extended period.

The standard for total hydrocarbons in drinking water was 10 μg/l but the basis was not clear. The evidence supported a safe level of 200 μg/l for infants and 300 μg/l for adults using a conservative assessment based on the molecular fractions present. A clearance level of 50 μg/l and/or no odour was determined to maintain a reasonably practical level but taking into account the danger of taste and odour problems. The objective was to clear individual zones but the problem was slow to clear. An overnight snapshot on the 15th/16th December showed that only two zones still had problem levels. However, demand was normal, in spite of the do not use instruction, and in order to avoid a surge in already normal demand resulting in excessive demand when the all clear to use water was given, the zones had to be released one at a time.

One of the lessons was that although the analysis of total hydrocarbons was useful it needed to be backed up with more specific analysis in order to determine the substances present in the mixture for an appropriate risk assessment but also to ensure correct interpretation of the data. An example of the importance of this was that the overnight assessment involved many hydrant samples and there were a number of high total hydrocarbon results. However, these were shown to be due to fluoranthene from particles in the hydrants and were nothing to do with the incident.

4 COMPLICATING FACTORS

There were a number of complicating factors in this incident including:

- the relatively new organisation

- the region had been recently through a major *E.coli* O157 outbreak that dictated considerable caution by the public health authorities.

- that plant could be brought on site without management's knowledge and the lack of proper procedures by Scottish Hydro

- operating procedures were not sufficiently well established so that management did not know there had been a diesel spill until nearly 24 hours later

- lack of knowledge about the design of drainage on the works was an important delaying factor

- the high chlorine residual provided a distraction

- the fact that the smell in the works was interpreted as fumes from the generator rather than diesel in the filters served to delay recognition of the problem.

5 CONCLUSIONS

The deficiency in operating procedures and their documentation on both the part of WoSW and Scottish Hydro were probably the primary factors in the incident. However, as in all such incidents it was a combination of events that led to the initial situation developing into a major incident. In the author's experience any significant incident results from such a combination of factors that are obvious in hindsight but may be less obvious at the time. In particular there is usually information which is open to misinterpretation and which is often a key factor in an incident moving beyond the point at which it can be contained without significant contamination of public supplies.

Major lessons have been learnt including:

•Planning and risk assessment. New approaches will help greatly. Multi-agency involvement.
•Need clear operational procedures.
•Always a combination of factors that turn an incident into an emergency.
•Information requirements significant.
•Basis of standards needs to be clear.
•Losing a supply carries significant health risks.

The developing approaches for managing microbial and chemical hazards from source to tap will contribute to a significantly reduced likelihood that such an incident could happen again. The new approach to adopting a more proactive risk management approach to drinking water quality from catchment to tap as outlined in the Bonn Framework will help the industry to prepared for and protect against such incidents in the future.

PROBLEMS, PERCEPTIONS AND PERFECTION – THE ROLE OF THE DRINKING WATER INSPECTORATE IN WATER QUALITY INCIDENTS AND EMERGENCIES

Claire Jackson
Deputy Chief Inspector, Drinking Water Inspectorate

1   INTRODUCTION

How do we define a water quality problem? Is it:
- an event that may have an impact on water quality; or
- an event that that has already impacted on water quality?

By water quality, do we mean:
- the aquatic environment in general; or
- a specific water body, treated or untreated?

At what point does a problem become an incident or a full blown emergency?
And how significant must the problem be before a consequence is perceived?
Fish kills and environmental damage may be part of that perception; as may be consumer reaction to discoloured water or water with an unusual taste or odour. The water does not actually have to be harmful to health to create major concerns and the resulting loss of public confidence in water supplies can itself lead to significant problems.
The dictionary definition of perfection includes:
- making perfect;
- faultlessness; and
- comparative excellence

The latter is something that we may all aim for; it is something that the regulators expect of the industry, but can we hope to attain perfection when dealing with an emergency?
This paper looks at the role of the Drinking Water Inspectorate in investigating incidents and emergencies affecting drinking water quality and the associated problems and perceptions.

2   THE ROLE OF THE DRINKING WATER INSPECTORATE

The Drinking Water Inspectorate acts for and on behalf of the Secretary of State and the National Assembly for Wales to ensure that water companies in England and Wales meet their regulatory obligations in terms of drinking water quality.   In this capacity the Drinking Water Inspectorate has a technical audit role for public water supplies, with powers of enforcement and prosecution.

Under Section 68 of the Water Industry Act 1991, water companies have a duty to supply wholesome water for domestic and food production purposes. However events occasionally occur that might impact on the quality or sufficiency of the water supplied. Water companies are required to notify such events to the Drinking Water Inspectorate under the terms set out in the Water Undertakers (Information) Direction 1998. This duty is enforceable under Section 202 of the Act.

The relevant section of the current Information Direction requires water companies to notify the Inspectorate of :

- the occurrence of any event which, by reason of its effect or likely effect on the quality or sufficiency of water supplied by it, gives rise or is likely to give rise to a significant risk to the health of the persons to whom the water is supplied. This will include any event notified by a water undertaker to a local or health authority under regulation 30(5) of the Water Supply (Water Quality) Regulations 1989.

- any other matter relating to the supply of water which:
  - o   in the opinion of the undertaker is of national significance; or
  - o   has attracted or, in the opinion of the undertaker, is likely to attract significant local or national publicity; or
  - o   has caused or, in the opinion of the undertaker, is likely to cause significant concern to persons to whom the water is supplied.

- any reports of disease in the community which it appears might possibly be associated with a water supply.

These criteria apply only to public water supplies.   The responsibility for monitoring private water supplies rests with Local Authorities.

The wording of the current Information Direction, and its predecessor Directions, deliberately leaves the decision to the water companies on what should be notified to the DWI. This is because an event that appears significant to a small company could appear to be less so to a larger company. Furthermore the trigger for consumer complaints may be very different between a rural area and a highly populated inner city area. Given these perceived anomalies, the Inspectorate has, over the years, issued guidance to the industry on the type of events that it considers should be notified. The most recent guidance was given in 1999 and included a long list of the sort of events that should be notified, with the caveat that this was not definitive. The same Information Letter (13/99) also provided guidance on the investigation process carried out by the Inspectorate.

3   NOTIFICATION OF WATER QUALITY EVENTS

The Information Direction requires water companies to notify the Inspectorate of an event as soon as maybe, usually by way of a telephone call. The company must then provide an initial report in writing, within 72 hours of the notification. This report should contain details of the event, including the cause, if known, and the actions taken and being taken by the company to deal with the impact and return supplies to normal.

The Inspectorate then assesses the information provided to determine whether the event meets the criteria of an incident, as defined by the Inspectorate, namely:

- a non trivial or unexpected breach of Part II of the Water Supply (Water Quality) Regulations 1989, as amended; or
- a breach of Part IV of the 1989 Regulations; or
- an unusual deterioration in water quality; or
- a significant risk to the health of consumers; or
- significant consumer perceived adverse water quality changes; or
- significant local or national media on a water quality issue that could result in consumer concern.

Occasionally further information is needed before a decision can be reached but, if the event is deemed to be a non incident, the water company is informed and thanked for making the notification. If the event is deemed to be an incident, the company is required to submit a full report within 30 days of the notification. Occasionally an extension is granted to the reporting period, particularly if full investigations into the incident cannot be completed within the timeframe or if external reports have to be included.

4   NUMBERS OF NOTIFICATIONS

Between 1990 and 2001 there was a steady year on year increase in the number of notifications received. This was due to the industry becoming more familiar with the process and the type of events that should be notified. In recent years some 60-70% of notifications have been classified as non-incidents. This does not mean that there are not lessons to be learnt, but rather that the industry is responding pro-actively to perceived problems before they develop into full blown incidents. The Inspectorate welcomes this approach because we can then alert other key players such as our policy colleagues, press office, and the Food Standards Agency of a developing situation.

Very occasionally we are made aware of a water quality problem by a third party. For example, a health authority may alert us to an increased number of reported cases of cryptosporidiosis at the same time as informing the water company. This is acceptable. However it is not acceptable for us to be made aware of a problem by irate consumers.

The table below shows the number of notifications received between 1995 and 2002. It also shows the number that were classified as incidents and the number of cases taken forward for prosecution.

Table 1. Number of notifications received between 1995 and 2002.

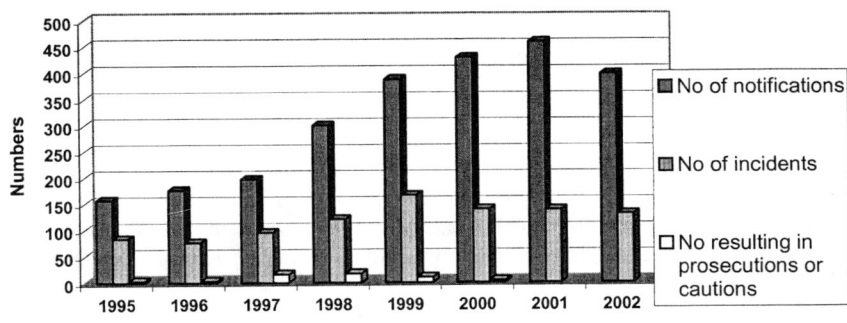

In the early years, many of the incidents related to bacteriological failures or problems at water treatment works. But by 1997, 33% of incidents related to the supply of discoloured water and this increased to more than 60% in 1998. The Inspectorate considered that many of these incidents were avoidable and a tough line was taken. Since 1998 the number of discoloured water incidents has gradually decreased, as companies have responded with improved planning and better operational management.

5 INVESTIGATIONS OF WATER QUALITY INCIDENTS

Most incidents are relatively minor happenings. All are assessed thoroughly and often result in recommendations to the company concerned on the actions needed to minimise the risk of future failures. Where the lessons to be learnt might benefit other companies, generic guidance may be issued to the industry.

Consideration is given to whether the company contravened any of the wholesomeness standards set out in the Water Supply (Water Quality) Regulations 1989 during the incident. We also consider whether the company contravened any other enforceable regulatory duty. If such contraventions occurred, the Inspectorate then has to decide whether the breaches were trivial or likely to recur, and whether enforcement action under Section 18 of the Water Industry Act 1991 is required.

The assessment also takes into account of the actions taken by the company to protect consumers and whether the company followed the advice given in 'Guidance on

Safeguarding the Quality of Public Water Supplies', and whether it followed industry best practice.

Depending on the circumstances, we may have to consider whether water unfit for human consumption was supplied during the incident and, if so, whether the company took all reasonable steps and acted with all due diligence to prevent the incident occurring. If there is sufficient evidence to show that water unfit for human consumption was supplied, then prosecution under Section 70 of the Water Industry Act may be considered.

6 EMERGENCIES

At what point does an incident become any emergency? Is it the number of consumers affected, the level of risk to public health caused by type of contaminant, the severity, the period that consumers are likely to be without drinking water or a combinations of these and other factors?

Water companies are used to dealing with straight forward incidents such as treatment failures; loss of supplies due to burst mains or service reservoirs going dry; planned work going wrong; microbiological failures; and unusual tastes/odours. They have tried and tested procedures in place for dealing with the cause and effect, for sampling and analysis, and for providing alternative supplies, as necessary. However a major incident for a large company, with adequate resources to deal with it, may well be considered an emergency by a small company with limited resources.

Section 208 of the Water Industry Act defines a civil emergency as:

'any natural disaster or other emergency which, in the opinion of the Secretary of State is, or may be likely, in relation to any area –

(a) so to disrupt water supplies or sewerage services; or

(b) to involve such destruction of or damage to life or property in that area,

as seriously and adversely to affect all the inhabitants of that area, or a substantial number of them, whether by depriving them of any of the essential of life or otherwise.'

Thus an emergency is much more serious than the range of events that have to notified under the Water Undertakers (Information) Direction and it may well meet some or all of the criteria used to define an incident.

Fortunately large scale emergencies or disasters, which affect drinking water supplies tend to be the exception rather than the norm. The Inspectorate would most certainly be involved, but the police would normally take the lead in responding or, in the case of a national emergency, a nominated Government department.

The Inspectorate's regulatory role would remain. However it is unlikely that enforcement action or prosecution would be considered for accidental or deliberate contamination of water supplies by third parties, unless the subsequent investigations identified major failings in the company's preparedness and actions taken. It is also unlikely that the Inspectorate would become involved in the local operational management of the emergency. The water companies are best placed and best qualified to deal with these aspects. Our role would be to provide technical and scientific advice to the central group, set up to manage the emergency. We would also work very closely with our Defra colleagues, providing technical and other advice and contributing to Ministerial briefings and press releases.

Our main input would be during the emergency phase, when we would liase closely with the water companies in the affected area to determine the nature and concentration of any contaminant and provide advice, if appropriate, on analysis. We would also liase with the water companies and the Environment Agency on containment, treatment, and disposal of any contaminated water.

We would look to the water companies to restore supplies to normal within the shortest possible timescale, subject to the overriding need to protect life and public health, and to meet their regulatory responsibilities at all times. However emergency relaxations of standards may be authorised, subject to public health constraints. We would also expect the water companies to provide, if necessary, safe alternative supplies of drinking water until the emergency is declared closed.

Once the emergency situation has been resolved, we would investigate the circumstances and, if necessary, issue advice to the industry on any lessons learnt in terms of preventing a recurrence or improving the response.

7   CONCLUSION

The Inspectorate's regulatory role must never be compromised, whether investigating routine incidents or dealing with major emergencies. Thus there are three other words beginning with 'P' that the water industry needs to consider, namely:

Planning – whether it be carrying out detailed risk assessments before undertaking work on the distribution system, or planning and rehearsing for major emergencies. There is a need to think the unthinkable and to work out how to deal with it.

Preparation - this includes ensuring that there are suitable laboratory facilities available to deal with unknown substances, whether they be chemical, microbiological or radiological; having appropriate expertise available to assess the results of analyses and predict their impact on public health; providing sufficient quantities of safe alternative supplies of drinking water, as needed; and coping with the unexpected, whilst maintaining public confidence.

Performance - carrying out critical internal assessments immediately after all major incidents to identify areas where the company under performed and where improvements can be made. Do not wait for the Inspectorate's assessment, with its recommendations and the potential for enforcement action or possible prosecution for supplying water unfit for human consumption!

We can never predict the unpredictable but, with good planning and preparation, something approaching perfection may be achieved by way of a response.

DEALING WITH COMPANY, PUBLIC AND MEDIA PERCEPTION

M Scott
Process Measurement Technology Ltd

1 JUSTIFICATION

The writer's justification for making a presentation on this subject is based on more than forty years in the measurement and control industry, most of which in the last decade has been on environmentally sensitive topics. Secretary & Director of SWIG (www.SWIG.org.uk), co-ordinator of MuttE (Foresight LINK Award project) and co-ordinator of Water-Monet (EPSRC Network www.Water-Monet.org) and membership of the local liaison group with Rugby Cement on the burning of waste and as a fuel have all provided valuable insights on how those who have no knowledge or interest in measurement perceive pollution issues.

2 PERCEPTION

The hardest thing for the technically or scientifically competent to learn is that the facts are rarely important to the receiver(s) of the information but are essential for the credibility of the disseminator of the information. The perception of the receiver will completely dominate the way in which the picture of the incident or proposal is formed. Perception rules.

Spin is rightly discredited but how and when information or data is presented is as important as the information itself. The how and when will be different depending on the particular recipients and when the information is being disseminated. Therefore it is important to be clear who is telling what to whom and when. The generators of data and hence information may come from:

- Main Board
- Finance
- Public Relations
- Senior management
- Local Managers
- Operators
- Maintenance
- Etc.

The receivers of the data and information may include:

- Local community
- Local media
- National media
- Regulators
- Pressure Groups
- Knowledgeable individuals
- Etc.

There is clear scope for misunderstandings, misconceptions and inconsistency since anyone of the first list might have communication with anyone from the second list. It is well known that the public and the media are most likely to accept their input from the loudest and most colourful voices. The media have the same time and selling pressures as those who are trying to project the right message and the general public do not have any pressure on them to properly understand or analyse the information. If it is an incident that is being dealt with the regaining of trust must be seen to be a priority and honest transparent engagement with the "watch dogs" should be the approach. The following rules seem obvious but when an incident is being dealt with, by a wide range of departments with different skills, it is very easy to get confused and inconsistent.

- Establish a team leader and a clear communication structure
- Timing must be appropriate
- The message must be structured for the particular audience
- The message must be consistent
- The audience must intuitively trust the messenger.

The last point is a real challenge since it is possible that the most senior manager or a member of the public relations department might not be the best communicator when an incident needs to be explained.

3 MEASUREMENT DATA

In the context of this conference the message will be clear but what is often less clear is that the quality of the message will be dependent on measurement data which might be very suspect; even if data is actually available. The measurement data, or lack of it will contribute to the information received by a wide range of people/organisations most of whom will have little understanding or interest in the source of the data but may take far reaching business or personal decisions based on it.

It is a valuable and discomforting experience to attempt understand the message from the point of view of the recipients and with the recipients understanding of the issues.

It is usually overlooked that measurement is multi-discipline and multi-cultural with issues such as the following to be considered:

- Measurement device/methodology design
- Business support
- Application design

- Data transmission
- Data use
- Data storage
- Standards/Best practice support

UKWIR Report 00/PC/03/01 On-line instrumentation standards & practices concluded that there is wide spread mistrust of measurement data, partly as a result of the interactions arising from the above list. Pre-normative research for ISO TC147 and an EU Report by Anders Lynggaard Jensen concluded that comparability of data between member states is not possible. Sparse and disparate data must be merged to give a useable picture of an event or proposal and the users of the resulting information must have a comfortable feel for the quality of the underlying data and some idea where improvements might be made. All data has an uncertainty and non-measurement specialists are increasingly aware of this thorny issue so it must be creatively addressed.

4   WHO GETS WHAT?

As already noted there will be a wide variety of receivers of information and this is particularly true in the media. It is good internal discipline as well as essential for the external world to prioritise who needs to receive and retransmit what sort of information:

- Senior management
- Middle management
- Internal Departments
- Technical specialists
- Multitudinous external contacts

It is important to examine the level of detail needed for each intended recipient of the message and make the individual messages clear and complete in themselves. It is quite a good idea to consider that if the disseminator cannot describe the problem on the back of a visiting card then some one else should be found to do the communication. It is vital to hear the questions correctly and to answer the question where at all possible since intelligent listening instils confidence. Where the question is misguided or incorrect supplementary information can only assist if presented in the context of the question.

The preceding discussion implies that the "problem" is all that needs to be dealt with but the media, pressure groups and local communities are likely to expand the discussion to include other issues such as blame, compensation and retribution. There will also be those who want to cover other complaints and grievances not associated with incident or proposal being dealt with.

5   OVERALL

It seems very obvious to assert that the brief must be mastered, by those dealing with the "outside world", but it is all too easy in a possibly frenetic environment to take shortcuts or to answer non-related questions. Mastering the brief will help consistency of message and will require co-ordinate with colleagues; an essential element of good external communication.

6 CONCLUSION

There probably won't be a conclusion to the need for communication on an incident.

Every problem is different and the story may continue for decades or may just dribble away only to resurface unexpectedly many years later.

It has already been said spin is counter productive but the message must be presented appropriately to each type of recipient and the content will depend on timing. For a major incident it will be impossible be right all the time so the final suggestion is that no matter what - Don't get angry.

WATER RELATED CHEMICAL INCIDENTS AND THEIR TOXICOLOGICAL MANAGEMENT

V.S.G. Murray[1] and J. Gray[2]

[1] Division of Chemical Hazards and Poisons (London), Health Protection Agency, Guy's and St Thomas Hospital NHS Trust, Avonley Road, London SE14 5ER
[2] Drinking Water Inspectorate, Ashdown House, 123 Victoria Street, London SW1E 6DE

1 INTRODUCTION

This paper provides a summary of the work of the Chemical Incident Response Service (CIRS) and the role of the Drinking Water Inspectorate (DWI) on water related chemical incidents. CIRS joined the Division of Chemical Hazards and Poisons (London) (DCHP (London)), Health Protection Agency on April 1 2003. This paper reviews eight recent water related chemical incidents and provides a commentary on the 'Water contamination emergencies: can we cope?' conference's concerns of:
  -Analytical support
  -Health risks and toxicological advice
  -Interagency assistance
  -Communication issues

These examples show some of the issues relating to how the management of water related chemical incidents can minimise harm to the population. The importance of effective and rapid communications between all relevant bodies in the event of a chemical contamination incident affecting drinking water supplies and the ready availability of an assessment of potential health risks and provision of toxicological advice is highlighted. The collaboration between the bodies such as DCHP (London) and DWI offers an example of a health protection process that should be shared with other relevant bodies.

2 CHEMICAL INCIDENT RESPONSE SERVICE

2.1 Background to the Chemical Incident Response Service (CIRS)

The Chemical Incident Response Service (CIRS) was formed in 1995 as a specialised department at the Medical Toxicology Unit, Guy's and St Thomas' Hospital Trust following a research programme set up after a meeting at the Royal Society of Medicine in 1989[1]. CIRS was one of five Regional Service Provider Units (RSPUs) across the UK, that provide advice to health authorities on public health, environmental, scientific, toxicological and epidemiological aspects of chemical incidents. The area served by CIRS is shown in figure 1.

Figure 1: Area served by the Chemical Incident Response Service / Division of Chemical Hazards and Poisons (London) 1995-2003 © DoctorMap (Maps produced by Richard Mohan)

☐ Not DCHP (London)
▨ DCHP (London)

Health authorities, now Primary Care Trusts and Health Protection Units, have had a statutory responsibility for the protection of public health from environmental hazards. In a chapter specifically on chemical incidents, the NHS Guidance 'Planning for Major Incidents', states that the health authority 'must ensure that satisfactory arrangements are in place for handling the public health and health care aspect of the response to chemical incidents[2]. The current system, which requires public health doctors within health authorities to manage chemical incidents, has been in place since 1993. Experience of public health professionals in chemical incident management has been variable and relatively limited. Consequently, there has been a need for procedures to guide public health professionals towards effective management of chemical incidents in order to protect public health. A series of handbooks for public health professionals, accident and emergency clinicians and environmental health practitioners has been developed along with a chemical incident handbook. [3,4,5,6] A further handbook on the environment and public health is in process of being published.[7]

The Health Protection Agency is a new national organisation for England and Wales, established on 1 April 2003.[8] It is dedicated to protecting people's health and reducing the impact of infectious diseases, chemical hazards, poisons and radiation hazards. It brings together the expertise of health and scientific professionals working in public health, communicable disease, emergency planning, infection control, laboratories, poisons, chemical, and radiation hazards.[9] The Division of Chemical Hazards and Poisons is part of the Agency and will provide comprehensive expert advice and support for chemical incidents across England and Wales. Such potential health threats might involve chemical fires, chemical contamination of the environment, or the deliberate release of chemicals and poisons.

2.2 Water related chemical incidents reported to CIRS

A project to review water related chemical incidents reported to CIRS was undertaken for the period from October 1997 until April 2001.[10] A total of 246 water-related chemical incidents were reported to CIRS. These incidents were defined as involving primary or secondary contamination of water, including surface waters, groundwater, marine water and drinking water. Not all the incidents that have been identified as involving water pollution have been categorised as water incidents under the type category used in the CIRS database. Figure 2 illustrates the incidents as documented by type.

Figure 2: Water-related chemical incidents as type reported

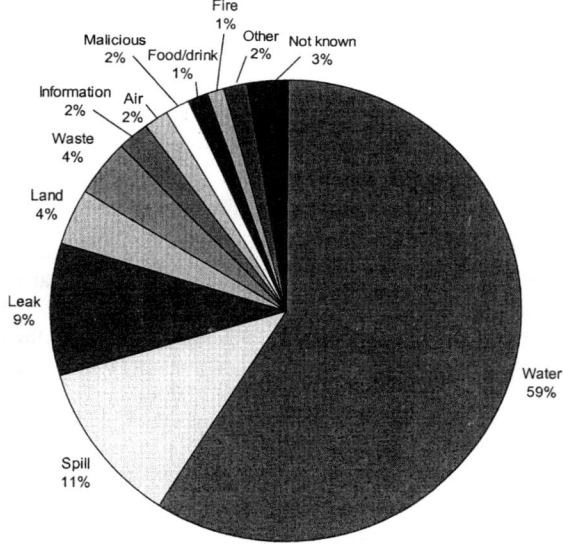

Figure 3 illustrates the most common chemicals involved in water-related chemical incidents. The most frequently occurring chemicals are hydrocarbons (including petrol and kerosene), iron, lead, nitrates, copper, cyanide, manganese and pesticides.

Figure 3a and b: Number of water-related chemical incidents by chemicals involved

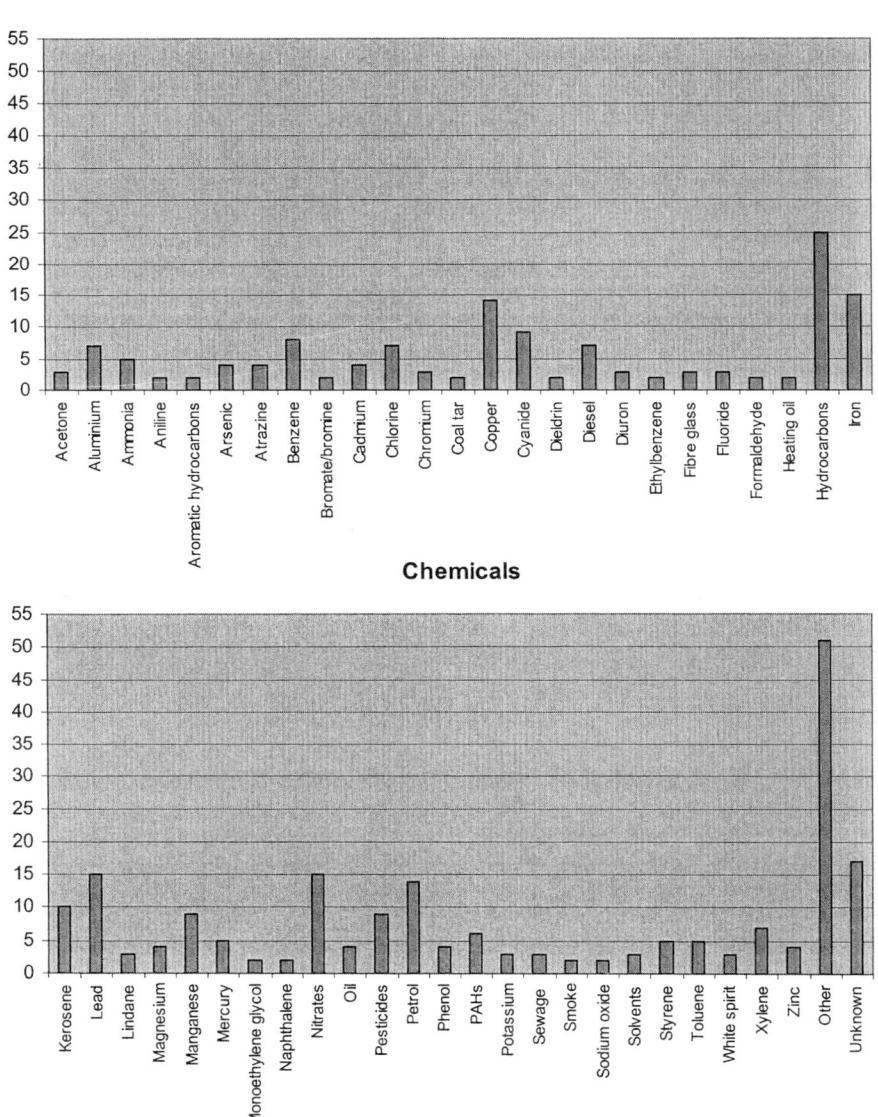

A line graph of CIRS incident type over the last five years (figure 4) shows the persistence of a considerable number of incidents involving release of chemicals to water.[11] Water incidents include releases to public and private water supplies as well as surface and costal waters. The lack of specificity in our classifications makes it difficult to compare this data with other agencies such as the Drinking Water Inspectorate and we consider improving our classification an important area of future work.

Figure 4: Trend of selected incident types over the period 1997-2002

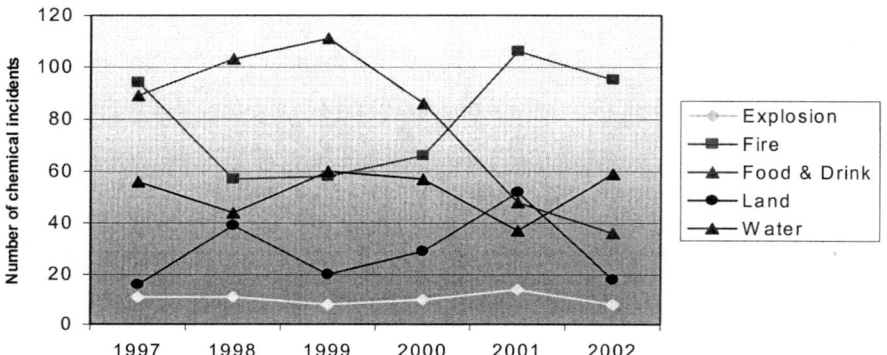

The key issues for chemical related drinking water contamination are the understanding of the potential toxicological and environmental health impacts. The Chemical Abstract Service reports that there are 5,401,571 commercially available chemicals.[12] Various other databases reporting that there are around 70,000 in common use. Unfortunately it is thought that reliable medical toxicology information for acute and chronic exposure is only available for 5,000 chemicals.[13] Not all these chemicals represent a risk to drinking water supplies. However many chemicals may not have standards against which it is possible to interpret the results of environmental investigations. For this reason a project was recently undertaken by CIRS to try and identify all web sites containing information on drinking water standards (Annex 1).

3 DRINKING WATER INSPECTORATE

3.1 Background to the Drinking Water Inspectorate (DWI)

DWI was established on 2 January 1990 following privatisation of the water industry. It comprised a nucleus of former staff from the Department of the Environment supplemented by others recruited from, among others, the water industry itself. DWI now operates within the Department for Environment, Food and Rural Affairs

3.2 Regulatory role of the Inspectorate

The Department for Environment, Food and Rural Affairs and the National Assembly for Wales are responsible under The Water Industry Act 1991 for regulating the quality of public drinking water supplies. These Authorities have appointed technical assessors, in the form of the Drinking Water Inspectorate, to act on their behalf. They have also delegated powers to the Chief Inspector to enforce water quality standards and initiate prosecutions.

Under Section 68 of the Water Industry Act 1991,[14] water companies have a duty to supply wholesome water for domestic and food production purposes. When events occur that might impact on the quality or sufficiency of the water supplied water companies are required to notify such events to DWI under the terms set out in the Water Undertakers (Information) Direction 1998. This duty is enforceable under Section 202 of the Act.

Specific details of what should be notified are presented elsewhere but include any event which gives rise or is likely to give rise to a significant risk to the health of the persons to whom the water is supplied and any reports of disease in the community which it appears might possibly be associated with the public water supply. The Inspectorate has issued guidance on the type of events that it considers should be notified, the most recent in 1999.[15]

3.3 Incident investigation

The Inspectorate assesses the information provided by water companies to determine whether the event meets the criteria of an incident, as defined by the Inspectorate, namely:

- a non trivial or unexpected breach of Part II of the Water Supply (Water Quality) Regulations 1989, as amended; or
- a breach of Part IV of the 1989 Regulations; or
- an unusual deterioration in water quality; or
- a significant risk to the health of consumers; or
- a significant number of consumers perceive adverse water quality changes; or
- significant local or national media interest on a water quality issue that could result in consumer concern.

Most incidents are relatively minor happenings but all are assessed thoroughly and may result in recommendations to the company concerned on the actions needed to minimise the risk of future failures. Where the lessons to be learnt might benefit other companies, generic guidance may be issued to the industry. Consideration is given to whether during the incident the company contravened any of the wholesomeness standards set out in the Water Supply (Water Quality) Regulations 1989.[16] DWI also considers whether the company contravened any other enforceable regulatory duty. If contraventions occurred, the Inspectorate then decides whether the breaches were trivial or likely to recur and whether enforcement action under Section 18 of the Water Industry Act 1991 is required.

The assessment also takes into account the actions taken by the company to protect consumers and whether the company followed the advice given in 'Guidance on Safeguarding the Quality of Public Water Supplies'[17] and whether it followed industry best practice.

Depending on the circumstances, DWI may have to consider whether water unfit for human consumption was supplied during the incident. If there is sufficient evidence to show that water unfit for human consumption was supplied, that the Company did not exercise all due diligence to prevent the incident from occurring and if it is in the public interest, then prosecution under Section 70 of the Water Industry Act may be considered.

3.4 Drinking water standards

Standards relating to the quality of drinking water supplies are contained in the Regulations. These Regulations are currently in a state of transition with most of the new quality standards, which relate to the 1998 EC Directive[18] coming into force on 25 December 2003. Until December 2003, the standards contained within the Water Supply (Water Quality) Regulations 1989 apply.

Standards are linked to the World Health Organisation guideline values for drinking water quality which are intended to protect public health as well as ensuring that water supplies are aesthetically acceptable to consumers. Under the EC Directive, standards will be subject to revision in the light of new knowledge. The new Regulations contain some new and revised standards but others, that are no longer appropriate, have been withdrawn.

3.5 Number of water related chemical incidents reported to DWI

Between 1990 and 2001 there was a steady year on year increase in the number of notifications received from water companies. This was attributed to the industry becoming more familiar with the process and the type of events that should be notified. In recent years some 60-70% of notifications have been classified as non-incidents. Occasionally DWI is made aware of a water quality problem by a third party, for example, a health authority may inform DWI of an increase in the number of reported cases of cryptosporidiosis at the same time as informing the water company.

Figure 5 shows the number of notifications received between 1995 and 2002. It also shows the number that were classified as incidents and the number of cases taken forward for prosecution.

Figure 5. Number of notifications made to DWI between 1995 and 2002

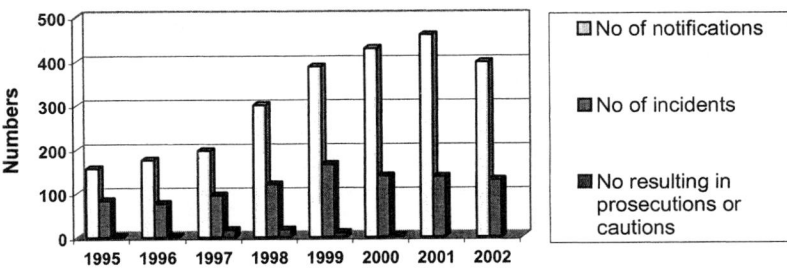

In earlier years, many of the incidents related to bacteriological failures or problems at water treatment works. In 1997, 33% of incidents related to the supply of discoloured water. This increased to more than 60% in 1998, which was attributed to the condition of the distribution systems and the associated remediation work being carried out. The Inspectorate considered that many of these incidents were avoidable. Since 1998 the number of discoloured water incidents has gradually decreased as companies have responded with improved planning and better operational management

4.   EXAMPLES OF CHEMCIAL WATER CONTAMINATION INCIDENTS

4.1 Examples from CIRS

4.1.1   Incident 1: On the Tuesday morning following a May Bank Holiday workers on a small industrial estate reported to a local water company a complaint about the taste and odour of the drinking water. This water supply was also used by the local housing estate residents. Trichloroethene was found in the drinking water supply at a level of 1080 mg/l. An alternate water supply was provided. The cause of the incident was a spill on industrial estate by a metal degreasing company. [19]

Commentary: On incident review by CIRS it was considered that analytical support was good; advice related to health risks and toxicology was adequate but interpretation was difficult with only a related WHO drinking water standard; it was felt that there was good interagency assistance; and that there was good communication between relevant bodies.

4.1.2   Incident 2: The area concerned a small housing development in a rural area. The residents lived in a recently converted stable block with oil-fired central heating. The first contamination event was reported to have been a leak of fuel contaminants into a lake in the grounds several years previously. A second contamination event with a leak from oil meters in the vicinity of water supply pipes near houses then occurred. Complaints about the water had been received for about a year. Remediation involved the installation of an overland water supply from the mains water supply. No adverse health effects were reported. The third contamination event occurred a year later with further complaints about odour and taste being reported. A local resident required dialysis. A questionnaire was sent to the residents to identify any adverse health effects. No adverse health effects were reported. Sampling of water and soil was undertaken to establish levels of exposure. Remediation involved the removal of contaminated soil and replacement of plastic plumbing. [20]

Commentary: On incident review by CIRS it was considered that analytical support was delayed; advice related to health risks and toxicology was adequate; it was felt that by the second and third events there was good interagency assistance and good communications between relevant bodies.

4.1.3   Incident 3: On April 1 2002 a water company reported that tests revealed a small quantity of diesel had entered the water supply. Local shops sold out of bottled water after the company's original advice not to drink or use water for cooking and local doctors (GPs) advised people to avoid tap water if it smelt. A local radio broadcast by the water company at approximately 12.00 hours on Tuesday 2 April provided further advice to the public. The first notification of any incident to the Chemical Incident Response Service by the local Health Protection Unit (HPU) occurred at 13.00 hours on 2 April. An incident meeting was called for 3 April and adverse health reports were reviewed along with enquiries and complaints to company. A cross-sectional survey was designed in collaboration with the local HPU and water company. Exposure was defined in two ways: first, by using information on measured hydrocarbons in drinking water to classify postcodes in 'more exposed' (up to 11-13µg/l) and 'less exposed' (up to 4-8µg/l), second, by asking about amount of tap water drunk. A postal questionnaire was sent out within a week, with a 35% response rate. This survey showed that people who drank more tap water reported more symptoms (a graded response) but this effect was confined to those respondents that were aware of the incident. People were also more likely to report drinking more water if they were aware that the pollution incident had occurred. A possible interpretation of this pattern is that people aware of the incident were more likely to recall the amount of water drunk and/or symptoms experienced, however an effect of hydrocarbon ingestion on gastroenteric symptoms cannot be excluded by this survey. The questionnaire invited respondents to add their own comments about the incident. These suggest that the public wants to know what is happening at the earliest opportunity and that surveillance of possible health effects appears to be reassuring rather than anxiety provoking. [21]

Commentary: On incident review by CIRS it was considered that analytical support was good; advice related to health risks and toxicology was adequate; but it was felt that there was some initial delay in interagency assistance and in communication between relevant bodies.

4.2 Examples from DWI

4.2.1   Incident 4: On day one, a consumer reported a taste in the water supply. Three affected properties were supplied with bottled water and the consumers were advised not to use the water. The water company sampled on day 2 and confirmed the presence of organic chemicals including 2(methylthio)benzothiazole, dimethyl butanedioic acid and bromohexanol at concentrations up to 0.5 $\mu g/l$. Unidentified hydrocarbons at concentrations up to 10 $\mu g/l$ were also detected. All three properties had been recently connected to a new medium density polyethylene main. The main was flushed and water quality returned to normal. The Company sought toxicological advice but advice was only available for related compounds, not for the specific compounds identified.

Commentary: On incident assessment by DWI it was considered that analytical support was good; advice related to health risks and toxicology was limited in that no specific toxicological information was available for the compounds in question with data only available on related compounds.

4.2.2 Incident 5: On day 1 a consumer reported a petrol-like taste in the water supply. The water company flushed the main and apparently resolved the problem. On day 14, the same consumer again reported a petrol/rubber cement taste in water supply. The Company sampled the affected property, an adjacent property and a hydrant on day 18. Analytical results were not available until day 29 which confirmed that the value for odour exceeded the standard. On day 32, a neighbour reported a benzene-like taste in the water supply. The company flushed the main the same day believing the problem to be associated with 'a build up of iron in the mains'. The presence of dissolved hydrocarbons up to a concentration of 31 $\mu g/l$ was confirmed on day 36. The Local Authority and Health Authority were advised and the latter sought toxicological advice. Bottled water was supplied to consumers in the seven properties affected. Investigations by the company revealed no obvious source of the contamination. The company decided to replace the MDPE communication pipe and during excavations discovered a layer of bituminous material some 0.5m thick close to one of the affected properties just under pavement level. This had an odour similar to that found in the consumers' supply and was subsequently shown to contain toluene, kerosene and diesel hydrocarbons.

Commentary: On incident assessment by DWI it was considered that analytical support was delayed, as was advice related to health risks and toxicology; there were also delays in establishing interagency assistance and in communications between relevant bodies.

4.2.3 Incident 6: On day 1 the manager at a water treatment works was informed that concentrations of isoproturon had been detected by on-line organic monitoring equipment at concentrations up to 4.3 $\mu g/l$ in the raw water source supplying the treatment works. Concentrations of isoproturon in the treated water were up to 1.3 $\mu g/l$. The dose of powdered activated carbon was immediately increased. Samples taken from the associated service reservoirs and some consumers' properties on days 1, 2 and 3 contained up to 1.5 $\mu g/l$ isoproturon in excess of the standard but less than the WHO guideline value. The company notified the Local Authority, Health Authority and the Environment Agency on day 2.

Commentary: On incident assessment by DWI it was considered that analytical support was good but that there were some delays in the communications between relevant bodies.

4.2.4   Incident   7:   On   what   was   to   become   day   one,   an   unknown   quantity   of chlorpyriphos   was   discharged   into   a   river,   6   km   upstream   of   the   abstraction   point supplying a water company's water treatment works.   The Environment Agency became aware of the problem on day 6 when invertebrate deaths occurred in the river.   On day 8 the Environment Agency notified the company which arranged for the analysis of the previous day's sample for chlorpyriphos and arranged for further samples of the raw water   to   be   taken.   A   sample   of   raw   water   taken   on   day   10   contained   0.325 $\mu g/l$ chlorpyriphos although none was found in the treated water leaving the works.   No samples were taken of treated water in supply.   The Local Authority, Health Authority and DWI were informed on day 12.

   Commentary: On incident assessment by DWI it was considered that analytical support was delayed and that there were also delays in interagency assistance and in communications between relevant bodies.

4.2.5   Incident   8:   Prior   to   coming   into   force   of   the   Water   Supply   (Water   Quality) Regulations 2000, the water company started monitoring in May 2000 its sources for compliance with the new standards, including bromate. It identified one ground water source with a concentration of > 100 $\mu g/l$ bromate which it confirmed by subsequent sampling. The source was isolated from supply and discussions initiated with the Health Authority, the Local authorities, DWI and the Environment Agency. Samples were taken from private supplies and other water sources. It was concluded that the contamination arose from a previous industrial site.   Monitoring of the plume continues in 2003 but remediation has not yet been initiated.   The company continues to manage the loss of an important and significant source of water.

   Commentary: On incident assessment by DWI it was considered that analytical support, assessment of health risks and provision of toxicological advice, interagency assistance and communications between relevant bodies were good.

5 CONCLUSION AND RECOMMENDATIONS

Over recent years many incidents of chemical contamination of water have been reported in England. Fortunately, so far, few have resulted in significant adverse health effects. However, experience in responding to these events has shown that to provide effective support to the public and consumers close links between public health organisations, water companies, the regulators and other related bodies are essential. This paper has addressed some of the common features between the perspectives of medical toxicology, public health, and drinking water quality regulation. From this collaboration it is possible to identify learning needs and, increasingly, the recognition of public accountability.

Annex 1: WEB ADDRESSES FOR WATER QUALITY STANDARDS. [22]

UNITED KINGDOM
Drinking Water Inspectorate
http://www.dwi.gov.uk
Main home page for the Drinking Water Inspectorate, from which specific searches can be carried out and direct links can be accessed to yield further information regarding Drinking Water Quality.

Water   Quality   Consultation   on   Regulations   for   Drinking   Water
http://www.defra.gov.uk/environment/consult/watersup/pdf/watersup.pdf

The water supply (water quality) regulations 2000
Stationery   Office   -   UK   Water   Quality   Legislation   2000
http://www.hmso.gov.uk/si/si2000/20003184.htm#Sch1p2
Full written legislation accessible in standard format detailing all of the UK Drinking Water standards

National Centre for Environmental Toxicology (NCET) http://www.ncet.co.uk Based at the Water Research Centre (WRc), a private company primarily funded by the water industry. Can provide information on the toxicology of chemicals in drinking water.

UNITED STATES of AMERICA
Environmental Protection Agency EPA - Office of Water http://www.epa.gov/ow/ Office of Science Technology - Department of the Office of Water http://www.epa.gov/waterscience/ Provides direct links to various issues concerning water quality, including standards and guidelines for chemical levels in drinking water.

Water quality criteria and standards http://www.epa.gov/ost/standards Extensive list of direct links to various topics concerned with the Water Quality Standards and Guidelines in the USA.

National Primary Drinking Water Regulations - EPA's Drinking Water Standards
http://www.epa.gov/safewater/mcl.html
Details all of the current Drinking Water Standards that are in use in the USA at the present time.

Drinking   Water   Regulations   and   Health   Advisories
http://www.epa.gov/ostwater/drinking/standards/summary.html
Provides further information regarding the USA drinking water health guidelines paying further attention to the health impacts of water quality.

CANADA
Health Canada Online
http://www.hc-sc.gc.ca/english/index.html

Provides a vast amount of information regarding public and environmental health impacts. Health Canada Online - Drinking Water Guidelines http://www.hc-sc.gc.ca/ehp/ehd/catalogue/general/iyh/dwguide.htm
Provides information on the standards and how they were derived, plus a direct link to a page where the Guidelines for Canadian Drinking Water Quality can be purchased-no PDF format available online.

OTHER WATER REGULATION WEBSITES

AUSTRALIA
The Cooperative Research Centre for Water Quality and Treatment – Australia
http://www.waterquality.crc.org.au

Australian Drinking Water Guidelines in PDF format (Needs to be purchased).
http://www.waterquality.crc.org.au/GUIDE.HTM

Information related to water quality with respect to biological, chemical, radiological and physical characteristics of drinking water. Further Information on the Australian Drinking Water Guidelines http://www.waterquality.crc.org.au/guidmore.htm Further information on the current drinking water guidelines, and how to purchase a copy. Rolling Revision of the Australian Drinking Water Guidelines

http://www.waterquality.crc.org.au/guideRR.htm
Information into the process employed in the amendment of the Australian Drinking Water Guidelines

NEW ZEALAND
Ministry of Health – New Zealand
www.moh.govt.nz/moh.nsf/

Drinking Water Standards for New Zealand 2000
www.moh.govt.nz/moh.nsf/ea6005dc347e7bd44c2566a40079ae6f/70727db605b9f56a4c25696400802887f
Full documentation of the Drinking Water Standards for New Zealand 2000 in PDF format. Also contains individual chapters as separate PDF files. Review of the 2000 Drinking Water Standards

www.moh.govt.nz/moh.nsf/c7ad5e032528c34c4c2566690076db9b/9c57904f727879eacc256bb100143184
Review of the 2000 Drinking Water Standards, with amended documents as PDF files.

WORLD HEALTH ORGANISATION
WHO Water, Sanitation and Health Department
http://www.who.int/water_sanitation_health/dwq/en/
Provides information on variety of issues relating to water quality Guidelines for Drinking Water Quality-Summary Tables

http://www.who.int/water_sanitation_health/GDWQ/Summary_tables/Sumtab.htm
Links to summary tables extracted from the "Guidelines for drinking-water quality, 2nd ed. Vol. 2 Health criteria and other supporting information, 1996 (pp. 940-949) and Addendum to Vol. 2 . 1998 (pp. 281-283) Geneva, World Health Organization." Includes information regarding acceptable bacterial, chemical and radioactive compound limits.

EUROPE
European Union – Environment Department including Water Quality Water Policy in the European Union http://www.europa.eu.int/comm/environment/water/index.html
Information regarding the water quality guidelines for the European Union, grouped according to the type of water source.

New Drinking Water Directive
http://www.europa.eu.int/comm/environment/water/water-drink/index_en.html
Information specifically regarding drinking water in Europe

Vivendi Water Systems- French Industrial and Municipal Water Treatment Company www.otv.fr/index_uk.htm Main home page for water information and products and services provided by the company. Guide de L'Eau http://www.guide-eau.com/ English and French versions. Page contains information regarding water and the environment and list of public and private companies involved in water remediation. Also includes a link to an extensive database.

KOREA
Water Korea
http://www.water.or.kr/engwater/ewk_main_bottom.html
Home Page for Water Korea, providing information regarding aspects of water in Korea and links to other sites. Water Quality
http://www.water.or.kr/engwater/ewk_menu_3.html

Water Quality Guidelines in use in Korea, including trends and examples of water quality incidents overseas. http://www.water.or.kr/engwater/quality/ewk_qul_st_drinking.html
Drinking water quality specifications for Korea

JAPAN
Japan for Sustainability – Water Laws http://www.japanfs.org/japan/laws.html#water
Provides links to pages detailing the laws governing waterways in Japan. Also contains links to other environmental laws.

National Institute of Health Services - Waterworks Law http://www.nihs.go.jp
Provides links to various issues regarding water safety and the laws in place to uphold any regulations
http://www.nihs.go.jp
Provides direct link to the Japanese drinking water quality standards

Acknowledgement
We would like to thank Adrienne Edkins and Amanda Welsh for their help in verifying these web sites on 7 May 2003, Nannerl Herriott for checking the epidemiology information from CIRS, Dr Giovanni Leonardi for additional information on incident 3 and Matthew Drinkwater for help in editing.

Particular thanks go to Dr Faith Goodfellow for her work on chemical contamination of water whilst completing her Engineering Doctorate – this was part of a four year doctoral research project funded by the Engineering and Physical Sciences Research Council (EPSRC) and CIRS. The research project was part of the University of Surrey and Brunel University Engineering Doctorate programme in Environmental Technology.

References

1   V.S.G. Murray, (editor) Major chemical disasters - medical aspects of management: Proceedings of a meeting arranged by the Section of Occupational Medicine of the Royal Society of Medicine, London 21-22 February 1989. International Congress and Symposium Series Number 155. Royal Society of Medicine: London, 1990.

2   NHS (1998) 'Chapter 8 - Chemical Incidents' in NHS (1998) *Planning for Major Incidents.* http://www.doh.gov.uk/epcu/nhsguidance.htm

3   D. Irvine, D. Cromie and V.S.G. Murray, *Chemical Incident Management for Public Health. Physicians* Chemical Incident Management Series. The Stationery Office, 1999. ISBN No: 011-322-017X

4   J. Fisher, D. Morgan-Jones, V.S.G. Murray, and G. Davies, *Chemical Incident Management: Accident and Emergency Clinicians.* Chemical Incident Management Series The Stationery Office, 1999. ISBN No: 0-11-322106-1

5   R. Fairman, V.S.G. Murray, Kirkwood and P. Saunders, *Chemical Incident Management for Local Authority Environmental Health Practitioners.* Chemical Incident Management Series. The Stationery Office. 2001 ISBN 0-11-322121-5

6   C. Farrow, H. Wheeler, N. Bates and V.S.G. Murray, *Chemical Incident Management Handbook*, Chemical Incident Management Series. The Stationery Office, 2000 ISBN 0-11-322252-1

7   E. Eagles, F.J.L. Goodfellow, F. Welch and V.S.G. Murray, *The Environment and Public Health.* Chemical Incident Management Series. The Stationery Office, in press

8   Chief Medical Officer, Getting Ahead of the Curve, Department of Health, England, 10 January 2002 http://www.doh.gov.uk/cmo/idstrategy/index.htm

9   Health Protection Agency web site for the Division of Chemicals and Poisons. http://hpa.org.uk/hpa/right_nav/chemicals.htm

10   F.J.L. Goodfellow, EngD thesis, University of Surrey, September 2001

11   G.S. Leonardi, N. Herriott, R. Mohan and N. Edwards, *Chemical Incident Report,* January 2003, No. 27, 2-13.

12   Chemical Abstracts Service, a division of the American Chemical Society http://www.cas.org/cgi-bin/regreport.pl

13   P.J. Baxter, BMJ, 1991 Jan 12, **302**, (6768), 61-2

14   HMSO, The Water Industry Act, 1991, 25 July 1991

15   DWI, Information Letter 13/99, 28 July 1999

16   HMSO, The Water Supply (Water Quality) Regulations 1989, July 1989

17   HMSO, Guidance on Safeguarding the quality of public water supplies, 1989

18   Council Directive 98/83/EC on the Quality of water intended for human consumption, 3 November 1998

19   F.J.L. Goodfellow, V.S.G. Murray and S.K. Ouki, Proceedings of the Engineering Doctorate in Environmental Technology Annual Conference 1999.

20   F.J.L. Goodfellow, S.K. Ouki and V.S.G. Murray, Journal Chartered Institution of Water and Environmental Management, 2002, **16,** 85-89

21   G. Lewendon, G.S. Leonardi, M. Mold and B. Guttridge, Submitted to Public Health

22   K. Sheridan, *Chemical Incident Report,* July 2001, No. 21, 24-27.

WATER AND PUBLIC HEALTH

D. Clapham

Environmental Health Department, Bradford Metropolitan District Council, City Hall, Channing Way, Bradford, West Yorkshire, BD1 1HY

1 WATER AND PUBLIC HEALTH

The spread of major epidemic diseases such as cholera and typhoid has been associated with drinking water since the mid-nineteenth century. Prior to this many people wondered what could be causing these diseases and there were two main factions; the contagionists, who believed that small animalcules spread illness and the miasmists, who believed that diseases arose spontaneously from accumulations of noxious material. Although the contagionists eventually won the argument, the miasmist theory held sway for many years and led to the removal and cleaning up of a great many nuisances and disgusting watercourses. Florence Nightingale was a noted miasmist as was Edwin Chadwick, the 'father of environmental health'.

In the 1850's, Dr John Snow (recently voted the greatest doctor of all times, followed closely by Hippocrates) reasoned that because of it effects on the alimentary tract, the cholera 'poison' must be introduced into the body via the mouth. He conjectured that water was the medium and carried out experiments to test his theory. His 'grand experiment' involved comparing illness rates amongst Londoners using two different water companies – one taking Thames water from the middle of the City (at the point where the effluent of 2 ¼ million people was discharged) and another used comparatively cleaner water taken from upstream. It was during this experiment that an outbreak of cholera began in the Soho area of London and the local parish committee asked Dr Snow for help. Because he suspected water, Snow looked at the five wells in the area. Four of them were visually contaminated and probably not very popular but in the fifth, the Broad Street pump, the water was clear. Dr Snow asked the committee to remove the handle of the pump, and by doing so he elegantly brought to an end a cholera outbreak, which had caused over 500 deaths. Another member of the Parish Inquiry Committee, the Reverend Henry Whitehead, was left to eventually discover that the cause was washed soiled nappies of a child that had died of cholera at 40 Broad Street.

Water has continued to be associated with disease since those times. In the UK, as in the other developed countries, the major waterborne diseases of cholera and typhoid have diminished as water treatment and disinfection have improved and an ever-increasing number of people have connected to mains water supplies. A study of the major outbreaks of waterborne disease in developed countries nowadays will show that the majority have been associated with unchlorinated or defectively chlorinated supplies.

2 PRIVATE WATER SUPPLIES

Private water supplies are those that are not provided by the water undertakers and are the springs, wells and boreholes found mainly in the rural areas of the country. Public supplies usually provide drinking water that is of uniformly good quality and which is getting better all the time. Private water supplies in contrast are often of dubious quality and will remain so for a considerable while. Even within this category, domestic supplies are five times more frequently contaminated than private commercial water sources and the smaller the supply is, the more likely it is to be contaminated.[1]

As with mains supplies, all private water supplies found to contain faecal coliforms can contain pathogens. In the UK, private water supply microbiological testing produces a general microbiological failure rate of about 30 per cent. Correctly organized sampling programmes, ones that take account of the effect of precipitation on water quality, are more likely to have failure rates of 70 or 80 per cent.[2,3,4] These studies highlight the fact that about 50 per cent of the negative results in the national sampling programme (that is the 'satisfactory' ones where no coliforms have been found in 100 ml of water) are likely to be unreliable (false negatives). Private water supplies can therefore be considered to be an inherently dangerous form of drinking water. In the UK, the risk of contracting a disease by drinking water from private rather than public supplies can be over fifty times.[5]

If they are so bad, why is the countryside not awash with ill people suffering the effects of drinking water from private water supplies? The reasons are generally high immunity levels following repeated illness and a general under reporting of intestinal disease. A study in the UK suggested that for every single case of infectious intestinal disease identified by the national surveillance system, another 1.4 were identified by laboratories and 6.2 faecal specimens were sent to laboratories for examination as disease was suspected. In addition, for this single identified case, a further 23 cases visited their doctors and another 136 cases occurred in the community, but did not visit their doctor or report it.[6] This under-reporting is thought to be even worse for illness from private water supplies. A study in 1996 found that people on private water supplies consulted their GP less often than those on mains supplies. Just to illustrate the problem of disease under-reporting in rural areas, during a mains water outbreak of gastrointestinal illness in a Scottish village, although a large number of people become ill (711 out of a population of 765) only a few went to the GP and they were spread between three separate practices. The outbreak of 93 per cent of the population went virtually un-noticed.[7]

A (the) major health improvement of the 20th century was the use of chlorine disinfection. The majority of private treatment systems do not use chlorination. The US approach to small supplies is more thorough and chlorination is mandatory in most states. However, the US has another important way of assessing potential problems. That is whether a supply is or is not 'under the direct influence of surface water'. Groundwater in the US is normally expected to come from a deep aquifer, free from microbiological contamination and not subject to sudden changes in turbidity, conductivity, temperature or pH. They are usually protected against any contamination from the surface and can be relied upon to provide reasonably safe water. Any other supply, where a continuously good quality of water cannot be guaranteed, is considered to be 'under the direct influence of surface water'.[8] It should not be assumed that boreholes are never under the influence and spring supplies always are. A borehole for example, may be in a karst limestone region and easily contaminated. The supply to a spring on the other hand, may be well protected both by surrounding impermeable layers of rock and its collection chamber construction (although that is less likely). Every state has a set of guidance rules to assist engineers or health officials in deciding whether a particular supply falls within this category.[9] There will, for example, be a requirement to test the raw water for faecal coliforms on a regular basis for a particular period of time. There will also be an expectation to test the system after heavy rain. It is suggested that the UK should adopt this procedure for all private water supplies so that their safety can be assured. This US approach is similar to the sanitary survey approach being suggested by most UK experts for inclusion into new British private water supplies legislation.

There is continual media fuss about the quality of mains drinking water in the UK. Although a greater risk to health, there does not appear to be the equivalent amount attention to private water supplies. This is probably because the consumers of water from private supplies are usually the people who have to spend the money to put them right, rather than passing the responsibility to a faceless corporation.

3 RISK ASSESSMENT

Some things seem to creating hectares of newsprint, whilst others receive virtually no attention. Compare the interest in genetically modified food or mobile phone masts with that of lead in drinking water. There has never been a recorded case of human illness with the first two, while lead has been known to be poisoning people since the Roman times of Vitruvius the Architect, two thousand years ago. Unfortunately, the potential for a substance to cause illness is not directly proportional to the attention it attracts from the public, the media or politicians. For various reasons, certain contaminants in drinking-water attract more than their fair share of interest and if you are involved in drinking-water safety and have to explain health-related information to people, you should have an understanding of the issues that motivate them and how this affects their perception of potential problems. Risk assessment by individuals is weighted against hazards that they have no direct control over. The public will consider a risk to be a thousand times greater, if they have no choice over whether they take it or not, such as the drinking of mains water. There may be an additional mental risk-reduction if expensive improvements are required.

The problems attracting attention at the moment include potential carcinogens such as pesticides and disinfection-by-products. Proof concerning the risks from these is scarce. It is easier to attract attention to these unknown effects, than the less attractive aspects of gastro-intestinal diseases. DDT for example is universally considered to be a terrible substance. A report by the American Council for Science and Health however has looked at the evidence and questioned whether it is as bad as is generally made out. They say that in Sri Lanka, DDT reduced malaria cases from 2.8 million in 1948 to only 17 in 1963. After spraying was stopped in 1964, malaria cases reached 2.5 million by 1969.[10] The same pattern was repeated in many other tropical regions of the world. The U.S. Agency for International Development said that malaria would have been 98 per cent eradicated had DDT continued to be used.[11] From 1960 to 1974 WHO screened 2,000 compounds for antimalarial insecticides. Only 30 were judged promising enough for field trials and none were as safe as DDT (insecticides such as malathion, which are much more toxic than DDT, were used instead). All of the substitutes were considerably more expensive than DDT.[12]

A 1978 National Cancer Institute report concluded—after two years of testing on several different strains of cancer-prone mice and rats—that DDT was not carcinogenic.[13] As for the DDT-caused eggshell thinning, it is unclear whether, in actual fact, it did. In 1998, researchers reported that thrush eggshells had been thinning at a steady rate 47 years before DDT hit the market, probably due to urban encroachment.[14] Some organochlorines have been shown to have weak estrogenic activity, but the amounts of naturally occurring estrogens in the environment dwarf the amounts of synthetic estrogens.[15] A recent article suggested that the ratio of natural to synthetic estrogens may be as much as 40,000,000 to 1.[16] In addition, Dr. Robert Golden of Environmental Risk Studies in Washington, DC, reviewed the research of numerous scientists and concluded that DDT had no significant estrogenic activity.[17] Recently, charges have been made associating DDT with breast cancer. In October 1997 a large study that found no evidence that exposure to DDT increases the risk of breast cancer.[18] In the accompanying editorial Dr. Steven Safe, a toxicologist at Texas A&M University, stated, "weakly estrogenic organochlorine compounds such as PCBs, DDT, and DDE are not a cause of breast cancer."[19]

4 MODERN PROBLEMS

As the background levels of the major nineteenth century diseases decline, the chances of their contaminating water supplies in the developed world will diminish. Modern outbreaks will be much more likely to be of unknown, viral or protozoan origin. Cryptosporidiosis is a major headache for water companies. The most famous *Cryptosporidium* incident was the Milwaukee outbreak, which was reported to have caused 403,000 cases and several deaths. Despite these numbers probably being a huge over-estimation, it was a large outbreak attracting much publicity. The cost has recently been calculated as US$ 96.2 million. Even when the over-estimation is accounted for, it cost US$ 57.2 million.[20]

Cryptosporidium is also a problem in private water supplies. The first UK study was in the Yorkshire area and fifteen separate sites were sampled ten times each. The aim was to show that contamination of private water supplies by *Cryptosporidium* was a reality, as this had never been proved before. Supplies were therefore selected that had particularly poor source protection and animals in the surrounding area. The water was also tested when the supplies were most likely to be contaminated, that is following rain during the autumn period. Nine of the fifteen supplies were found to contain *Cryptosporidium* (60 per cent).[21] In addition, eight of the supplies also contained another protozoan parasite, *Giardia lamblia* (53.3 per cent). 98.6 per cent of the samples showed some form of faecal contamination and the levels of faecal indicators were almost uniformly high with a faecal coliform geometric mean of 224.5 per 100 ml and a maximum of 6,000 per 100 ml. The study proved that private water supplies without reasonable source protection could be considered to be at risk when animals are present.

An extra problem for private water supplies is associated with campsites and holiday homes. Where people are constantly drinking contaminated water from private water supplies, they may become immune to *Cryptosporidium* and other diseases. They will only become ill when they there is a new animal in the field or a new illness appears in the herd. Campsites, etc. contain transient populations who do not have this immunity and are therefore more likely to become ill. The population may move on before the illness starts or will blame it on something else, such as different foodstuffs or "holiday tummy". This is exacerbated if only part of a group is affected (children drinking cordials diluted with raw water rather than adults who drink boiled water beverages). Even so, campsites are most often associated with outbreaks of waterborne illness.

5 EMERGENCY EVENTS

Many of the diseases that are thought to be terrorist material, such as anthrax and tularaemia are zoonotic and found in rural areas. Tularaemia is caused by *Francisella tularensis*, a virulent organism found throughout most of the Northern hemisphere. The infective dose is up to 10^8 organisms if taken orally, but much less (ten to fifty) if infection is acquired through the skin. It has been associated with waterborne outbreaks in Italy, Turkey and possibly Kosovo. It is a zoonotic disease spread by a wide range of rodents, lagomorphs (rabbits and hares) and birds; outbreaks have been associated with both hares and voles in Bosnia. It is also know as 'deer fly fever' and 'water-rat trappers disease'.[22]

Water supplies may be easier to access in rural areas. If there is to be an emergency of a terrorist nature it may also be trialed in rural areas. In the US, surveillance is moving towards getting vets and rural medical people to notify out of the ordinary occurrences. This might be useful in other parts of the world. After the incidence of anthrax in America, which followed the dreadful terrorist attacks on New York on 11th September 2001, the US Government identified $1.5 billion for basic research into 'select agents' that could be used by bioterrorists. This was an increase of five times over the previous amount. As well as anthrax, smallpox, botulism and plague, tularemia was considered as one of these Class A agents.[23]

6 THE DEVELOPING WORLD

In the developing world water related disease continues to be the major killer, particularly of children. The following is a brief description of the most damaging.

Dengue fever - this is one of the most common and widespread insect-borne infections in the world. About fifty million people are afflicted annually, with 24,000 deaths.[24] Symptoms vary from fever to fatal haemorrhaging (serious bleeding from blood vessels). About 5 per cent of people who get the disease die from it –these are mainly children. The main species of mosquito that is the vector for dengue fever is *Aedes aegypti*. *Aedes albopictus* also spreads the disease, but is less important. *Aedes* are the type of mosquito that breed in rain-filled waste cans, tyres, etc. This species also transmits yellow fever.

Viral Gastroenteritis - thought to cause ten million deaths per year and up to five billion cases of illness. It is the number one cause of acute gastroenteritis in every country in the world, with increasing numbers of outbreaks in the temperate regions of the world. In the past they have been difficult to identify in the laboratory and do not grow in culture cells, thus they are also known as the 'fastidious' viruses. How something that causes five billion people to become ill in one year can be called 'fastidious' I do not know.

Typhoid - this is another of the world's major waterborne diarrhoeal diseases. There are over twelve million cases of typhoid and paratyphoid ('para' means resembling) per year, causing approximately 600,000 deaths annually. It is particularly prevalent on the Indian sub-continent but people living in, and travellers to South America, West Africa and South Asia can also contract it. Typhoid got its name because it produces symptoms similar to typhus, although lice rather than water spread that disease. In 1837, William Gerhard of Philadelphia showed that typhoid was different to typhus and in 1880, Carl Eberth and Edwin Klebs finally identified the causative organism, *Salmonella typhi*.[25] Typhoid is therefore a salmonella infection and there are many other types of salmonella that will cause human disease

Shigellosis or Bacillary Dysentery - dysentery is a widely used general term that basically means serious, bloody diarrhoea (or bloody serious diarrhoea). Hippocrates first used the term and at that time it meant diarrhoea with mucus and blood in loose stools. There are two types of dysentery; amoebic – which is caused *by Entamoeba histolytica* and bacterial - which is caused by various forms of *Shigella*. In fact several organisms cause bacillary dysentery but *Shigella* is by far the most important. Shigellosis is a major worldwide disease that affects two million or more people a year of which about 650,000 will die because of it. An epidemic in Central America that began in 1968 caused 20,000 deaths from over 500,000 cases in a four-year period.[26] Although in the UK cases *of Shigella sonnei* infection have fallen to a fifty-three-year low, it is now showing signs of antibiotic drug resistance, which is a worry for the future.

Bacterial dysentery has a normal incubation period of two or three days, although it can take up to seven days. Man is the only host, although the organism will live for some time in fresh water. The infective dose is less than 200 organisms in healthy adults and patients can excrete 10^9 per gram of faeces, so it can be very infectious. It is spread by the faecal-oral route and is associated with poor hygiene practices leading to person to person spread, food poisoning and water where there are inadequate water treatment and sanitary facilities.

Amoebic dysentery is cited as infecting one tenth of the world population, or 500 million people. This disease is found throughout the world. It has been claimed that 40 million people develop intestinal disease or liver abscesses annually, with 40,000 of them dying from amoebiasis. Some claim it is the third leading cause of morbidity and mortality due to parasitic disease in humans (after malaria and schistosomiasis). Others have estimated that it is responsible for between fifty and one hundred thousand deaths every year. The numbers vary because of the different ways disease statistics are collected and so will never be completely accurate, but it is clear that amoebic dysentery is one of the world's major diseases and that it causes untold misery.

Schistosomiasis – this is a disease that affects over 200 million people in 74 countries. Twenty million suffer severe consequences. This disease has been known to cause illness in humans since the days of the ancient Egyptians. It affects over two hundred million people in seventy-four countries in the developing world and twenty million suffer severe consequences from the disease.[27] Again, this is a disease that affects a great proportion of the world's children and the very poor and about 80 per cent of those affected live in Africa. Having schistosomiasis can have a chronic, severely debilitating effect on the body. Problems include anaemia and impaired development, both mental and physical. Although it is not as fatal as many other tropical diseases, about 20,000 people a year will die from it. Cases are normally identified by blood in the urine or faeces. As with many diseases of hot countries, migration caused by poverty or war brings people with the disease to new areas. This has happened recently in Somalia and Djibouti. Another problem with schistosomiasis is that changes to rivers caused by damming them will spread the disease to new countries. Around the man-made Lake Volta in Ghana, for example, 90 per cent of the children are now affected.

Malaria – this disease causes more illness than any other. It is spread by the mosquito and the infective organism is called *Plasmodia*. It is a single-celled protozoan parasite with four separate species -*falciparum, vivax, malariae* and *ovale*. They all have different drug-resistance, severity and geographical distribution. The symptoms of malaria are severe fever, headache and nausea. The parasite invades the red blood cells causing damage to organs and anaemia. Those who survive the initial attack often experience serious relapses. *Plasmodium falciparum* is a particularly fatal species of the malaria-causing parasite. It accumulates in and damages the brain, causing comas and death. It can also cause kidney failure and obstruct the circulation by damaging blood cells and turning the blood to what is sometimes described as 'sludge'. Another name for malaria is black water fever.

The disease is thought to be responsible for 267 million cases and one to two million deaths every year. 75 per cent of these deaths are in children under five years old. This huge death toll causes untold misery year on year. Many of the prophylactic (protective) drugs are becoming less useful as resistance to them spreads and others are found to have severe side effects. There were successful eradication campaigns in the

1950s and 1960s using cheap persistent chemicals, but there has been a resurgence of the disease since their use was stopped (see above). In Africa the disease is growing at a fantastic rate. Very high rates of under-five mortality are found with the figure being quoted as over 90 per cent. It appears that in sub-Saharan Africa the infection rate is already so high that it does not appear to be possible for the situation to worsen.

Drancunculiasis – this is another disease that is endemic in Africa. The disease used to be common in India and the Middle East but is now confined to rural areas in thirteen countries in Africa, between the Sahara and the equator.[28] Worldwide numbers of those infected fell from over three million in1986 to 220,000 in 1994. This disabling condition is caused by the roundworm parasite *Dranculus medinensis*. Its intermediate host is the copecod water flea. Copecods are very small and can barely be seen in a glass of water if is held up to the light. People become infected when they drink stagnant, dirty water containing the water fleas. When they enter the stomach, gastric juices kill the fleas but release the larvae. The larvae burrow through the stomach wall and migrate around the body. The worms, which are about the thickness of a piece of spaghetti, grow to nearly a metre in length. They mate, the males die and the pregnant females migrate to the skin. People do not know they are infected until the worm actually reaches the surface of the skin, which can take up to a year. Here they form a blister that eventually ruptures. At this time there are other symptoms such as nausea, diarrhoea, vomiting and dizziness. Because the blister feels hot (another name for the creature is 'fiery serpent' because of this hot blister), people bathe in cold water to cool down. This causes the female to release millions of immature worms in a milky white fluid. The water fleas then eat them, which starts the process off again. The worms emerge mainly through the legs and feet but in 10 per cent of cases they may come out through the arms, genitalia, trunk or buttocks.

'A visiting Martian could be forgiven for asking why it is that on one part of the planet so much effort is devoted to the elimination of an unproven link between water quality and cancer, while elsewhere tens of thousands die daily from eradicable disease'.[29] It is also interesting to speculate the finance that would be available if terrorists were killing off millions of children under the age of five every year. This of course is the children killed by dirty drinking water and poor sanitation annually

7 THE FUTURE

Clean water and sanitation should be a right for everyone on the planet. We have traditional, well-tried technologies such as slow sand filters that are known to eradicate disease when allied to the provision of sanitation and suitable health education. Whilst these should be provided as widely as possible, new technologies are being introduced for small scale and rural communities. These new technologies are low tech – filling clear plastic bottles with water and leaving them on roofs reduces bacteria due to the effects of UV light, time and settlement; protecting water once it has been collected in individual homes reduces gastrointestinal illness, fog curtains collecting water in desert mountainous regions produces clean water provided it is away from towns with high atmospheric lead levels and using old saris to filter water will remove the copecods associated with drancunculiasis, suspended particles and the bacteria that adhere to them. Technology is not the problem; it is the lack of money being allocated to it and investment in public health infrastructure.

References

1   M. Rutter, G.L. Nichols, A. Swan and J. de Louivois, *Epidemiological Infection,*
     2000, **124**, 417-425.
2   J. Barraclough, R. Collinge, and N.J. Horan, *Environmental Health,* 1988, **96**, (11),
     12-16.
3   T.J. Humphrey and J.G. Cruickshank, *Community Medicine,* 1985, **7**, 43 - 47.
4   R.A.E. Barrell, *Environmental Health,* 1989, **97**, (7), 171-173.
5   D. Clapham, MSc Thesis, University of Leeds, 1993..
6   S. Handysides, *Communicable Disease and Public Health,* 1999, **2**, (2), 78-79.
7   I.G. Jones and M. Roworth, *Public Health,*1996, **110**, 277-282.
8   R. Foster, Personal communication, 2003.
9   TDEC, *Guidance for Determining if a Ground Water Source is Under the Direct
     Influence of Surface Water* Tennessee Department of Environment and
     Conservation, Division of Water Supply, August 1991.
10   E.H. Gerberg and H. Wilcox III, *Environmental Assessment of Malaria and
     Control Projects–Sri Lanka.* Agency for International Development; 1977, **20**, 32–
     33.
11   J. Bast, P. Hill and R. Rue, *Eco-Sanity: A Common Sense Guide to
     Environmentalism.* Lanham, MD: The Heartland Institute; 1994, pp 100–101.
12   M. W. Service. *Ecological Effects of Pesticides* (Perring, FH and Mellanby K,
     eds.). New York: Academic Press; 1977, p.156.
13   E. Efron,*The Apocalyptics.* New York: Touchstone/Simon & Schuster; 1985, p. 268
14   S. Milius, *Science News. 1988,* **153**, (17), 261.
15   T.H. Jukes, *J Amer Coll Toxicol.* 1983, **2**, (3),147–160.
16   S. Safe, *Environ Health Perspect*; 1995, **103**, 346–351.
17   R. Golden, *Proceedings of the International Environmental Conference,*
     Washington, DC. 1995.
18   D.J. Hunter, S.E. Hankinson, F. Laden, G. Colditz, J.E. Manson, W.C. Willett, F.E.
     Speizer and M.S. Wolff, *N Engl J Med.* 1997, **337**, 1253–1258.
19   S. Safe, *N Engl J Med.* 1997, **337**, 1303–1304.
20   P.S. Corso, M.H. Kramer, K.A. Blair, *et al.*, *Emerging Infectious Diseases,* 2003, **9** 4
     Online article http://www.cdc.gov/ncidod/eid/vol9no4/02-0417.htm
21   D. Clapham, *Institute of Biology Conference,* 1997, University of Warwick
     (Warwick).
22   P. Hunter, *Waterborne Disease: Epidemiology and Ecology'.* John Wiley and Sons
     (Chichester, New York, Weinheim, Brisbane, Singapore, Toronto), 1997.
23   D. MacKenzie, *New Scientist,* 2002, 9/2/2002, 8-10.
24   WHO (2002) *Dengue and Dengue Fever. Fact Sheet Number 117*, Revised April
     2002.   http://www.who.int/inf-fs/en/fact117.html, World Health Organization
     (Geneva), 2002.
25   Anon, *Travel Health CDR Weekly,* 4 July 2002.
26   WHO, Epidemic dysentery, *Fact Sheet Number 108*, World Health Organization
     (Geneva), 1996.
27   WHO, Shistosomiasis, *Fact Sheet Number 115*, World Health Organization
     (Geneva) 1996.
28   E. Ruiz-Tiben, D.R. Hopkins, T.K. Ruebush and R.L. Kaiser, *Emerging Infectious
     Diseases* 1995, **1**, 2 online article http://www.cdc.gov/ncidod/EID/pastcon.htm
29   R.F. Packham, *Journal of Institute of Water and Environmental Management,* 1990,
     4 (5) 484-488.

ENVIRONMENT AGENCY ANALYSIS OVERVIEW ON CONTAMINATION MONITORING

M. Cooper

National Laboratory Service

1 INTRODUCTION

The Environment Agency relies heavily on the National Laboratory Service (NLS) to monitor contamination – this is an Internal Business Unit of the Agency. This organisation has some 350 scientists operating from 5 laboratories in geographically different locations across England and Wales. Samples of water, solid or gas are taken from anywhere across England or Wales and stored in one of 92 sample depots. An internal courier system transports these samples throughout the night to arrive at one of the laboratories in the early hours of the next morning. The integrity of this supply chain is critical, it acts as a chain of custody in court cases where prosecutions may be necessary, and the samples have to be stored under suitable conditions at appropriate temperatures.

The NLS has a significant technological resource available across the laboratories, covering organics, metals, inorganics, microbiology and hand-held meters. There is currently a trend of increased organics analyses - providing a capability to monitor complex organics such as pharmaceuticals in the environment. Metals analysis continues to be important – with speciation becoming more relevant as time goes on. The trend is for a decreasing volume of inorganics analysis – with a shift towards hand-held (on-site) devices for this type of work and a general reduction in demand.

All analytical work in the NLS is performed to appropriate standards – this is defined by the Environment Agency to be ISO 17025 (UKAS). Valid analysis is needed for court cases – a crucial part of delivering a service to the environmental regulator of England and Wales. There are also robust Health and Safety and environmental systems in place – with more recently business performance monitoring systems.

The majority of work in the NLS is routine monitoring, which provides robust systems and processes that may then be applied when a contamination emergency arises. Such events typically involve some form of pollution, where the response time is crucial. These events can occur at any time, so the NLS provides a 24/7 service 365 days of the year. The fast production of analytical information is needed to minimise the impact of a pollution event. Furthermore, the link between sampling and analytical result (ie. the chain of custody) is critical. This information must stand up in court for a successful prosecution. Four examples of such events are given today.

It is important to take into account all available information. What on-site evidence is available? - raining or sunny – with on-site testing being performed where appropriate. All of this data is combined with the analytical data to produce information of what has occurred – it is the summing together these parts in the form of a jigsaw that is important in contamination monitoring events.

2 EXAMPLES

A synopsis of three water pollution incidents, and the role of the NLS, is described.

2.1 Chemical Spill in Doe Park, Denholme (near Bradford)

An unknown chemical enters a public reservoir at Doe Park, forming a heavy scum on the surface and a large fish kill. Environment Protection Officers (EPO) place a boom across the reservoir and put in place absorbent materials in an attempt to remove some of the pollution. A careful note of the surroundings is taken by the EPOs – no buildings or work of significance is noted. Meanwhile a sample of the pollutant is taken to NLS Leeds laboratory for examination.

The sample is diluted with an organic solvent (dichloromethane) and injected into a Gas Chromatograph – Mass Spectrometer (GC-MS). After careful examination of the data, four groups of compounds were identified – white spirit, propiconazole, terbiconazole and permethrin. The point of discharge into the reservoir was tracked back – this led to a timber yard. Investigation of this yard revealed a leaking storage tank, which contained Protim 415T. It was housed within a containment bund constructed from porous material! Protim 415T contains all of the compounds identified in the GC-MS analysis – permethrin in particular is known to be very toxic to aquatic life. Having identified the pollutant in the sample, it was important to measure how much was present. Using a UKAS approved method, 250ng/l of permethrin was quantified in the sample.

Also, dieldrin was found at trace levels in the reservoir samples and the storage tank. Before being banned dieldrin was used in Protim formulations and had in the past been stored in this tank!

The outcome of this investigation was.........
- 2 month clean up of the Doe Park reservoir (closed to public during clean up)
- approximately 14,000 litres of protim 415T had entered the reservoir
- the timber merchants were taken to court by the Agency and pleaded guilty and were fined £8000, paid £6700 costs and there were £80,000 clean-up costs

2.2 Foot and Mouth Epidemic

By April 2001 the foot and mouth epidemic had reached crisis stage in Wales. The decision to use the Epynt mountain range as one of the burial sites angered local opinion. After all, this was an area of outstanding natural beauty and free from foot and mouth. As the bulldozers set to work, members of the public obtained what they saw as 'evidence' that the surrounding environment was showing signs of contamination from the burial site. The main concerns of the residents were associated with public health and there was anxiety over damage to the environment.

The Environment Agency had to act fast. An extensive monitoring programme was set up to monitor contamination levels in ground and surface waters.

NLS Llanelli laboratory had to provide a rapid method of tracking the effects of pollutants from the burial site. An incident room was set up at the National Assembly form Wales and included Assembly civil servants at the highest level, MAFF and EA representatives. Analytical data had to be made available daily as a matter of urgency – for discussion and dissemination in daily releases to the media and public information updates provided.

The Environment Agency set up a monitoring programme to look at levels of contamination in ground water. The NLS provided a 24 hour emergency incident response : -

- Ammonia used as the pollution indicator
- Staff worked a rota system to provide 24 hour cover
- Expert analysts provided opinions and interpretations
- Broad range of determinands were analysed during the extensive monitoring programme.

The rivers ran red with blood – but where was this coming from? There was a forensic analysis requirement to match blood types and animal by-products in environmental water systems. A method was developed on the hoof (!) using liquid chromatography with fluorescence detection. This enabled prominent peaks to be identified as the globulins, whose retention time varied depending on the animal species. This method enabled blood to be matched with the river samples, allowing sheep, human and cow blood to be compared. This proved that the rivers were running red with sheep and cow blood.

2.3 Major Fish Kill

Following a report of a major fish mortality in a high quality watercourse (a key breeding site for otters) investigating staff at the scene relayed information direct to the NLS laboratory at Starcross, Exeter. Intelligence pointed to a (illegal) store of tributyl tin as the most likely source and it was on this basis that samples were collected and sent to the laboratory.

It was essential that results were produced with a minimum delay, both in terms of informing the management of the incident and ensuring the integrity of any evidence was preserved.

Two analytical sections co-ordinated their efforts to achieve this. The Metals section, using ICP-MS, worked as 'rangefinders' to obtain fast results for total tin. This information was then passed on to the Organics section who used it to estimate the correct dilutions for GC-MS analysis. Using this approach, precise identification and quantification with a minimum requirement for repeats ensured a rapid turnaround of results.

TriButyl Tin was confirmed both in the water column and within the tissues of fish mortalities (up to 8600μg/kg). The only practical solution was to undertake a large water transfer in order to dilute the compound to sub-lethal limits.

This case is a good example of the potential 'long term' resource impacts which often carry over from pollution 'incidents'. To this day, the NLS is still analysing samples from both the watercourse and of soils beneath the site of the chemical storage and laboratory staff have produced over 200 witness statements for the pending court case.

3 SUMMARY OF KEY FACTS

- The NLS is the key analytical provider for the Environment Agency in terms of aqueous, solid and microbiological analysis.
- The NLS provides a 24/7, 365 day emergency service to the Environment Agency through five nationally distributed laboratories, collecting from 92 sample storage depots.
- The NLS performs a forensic role during pollution incidents as well as providing expert witness statements.
- All methods are UKAS accredited (ISO 17025)
- The NLS processes some 400,000 samples and 3,000,000 determinands per annum.

CONTAMINATION MONITORING: SCREENING VS TARGETED ANALYSIS

A. Clark

Dstl, Porton Down, Salisbury

1 INTRODUCTION

Current analytical methods used in the water industry for compliance monitoring are not well suited to detection of water contamination during emergencies. This applies to early warning of harmful chemical releases as well as the presence of biological contamination (which can take days rather than hours or minutes to provide a definitive result). A case is therefore made for improved rapid methods of analysis to meet this important requirement. Consideration is given to the relative merits of targeted analysis versus screening techniques, with examples provided. One approach to targeted analysis of organic contaminants is to optimise the performance of chromatographic separation and identification techniques. An alternative is to screen using a spectroscopic method such as UV absorption or Surface Enhanced Raman Spectroscopy (SERS), both of which involve minimal sample handling and have potential for at site use. Other potentially useful techniques for rapid screening for chemical and microbiological contaminants are also listed, some of which are described in greater detail in subsequent chapters. In conclusion, it is argued that targeted and screening techniques both have a role to play in contamination monitoring; using targeted analysis with optimised methodology for identification, combined with screening techniques in order to prioritise suspect samples for targeted analysis.

It may at times be appropriate to operate both of these methods in parallel to meet tight deadlines involved in providing timely information to decision-makers. Selecting the required methodology and analytical capabilities will depend on a suitable risk assessment as well as on the nature of the particular emergency. Guidelines are offered on how to select and make best use of currently-available screening techniques. While confirmatory identification may be secondary to speed of detection, rapid methods need to be robust, and based on standard operating procedures with clearly defined reporting protocols to enable them to function in an emergency. In addition, analysis needs to be based on well-rehearsed procedures, preferably in regular use. Although significant contamination events may be a rarity, it is important to have effective and appropriate contingency plans in place. Other benefits can also accrue from adopting rapid methods including improvements in day-to-day operating efficiency and business resilience. This is being aided by new developments in analytical procedures and sensor technology.

2 CURRENT METHODS OF ANALYSIS

Methods of analysis in routine use within the water industry are geared primarily to compliance with UK Drinking Water Regulations.[1] This predicates use of validated "Blue Book" protocols which are based on well-established methodology and employing standard analytical equipment.[2,3] Capability and capacity for rapid analysis of water samples within the industry is limited as regards providing early warning of a contamination event prior to exposure of the population. While rapid screening of metal toxicants can be met by use of ICP-MS[4,5] the same cannot be said for organic chemicals or biological contaminants*.

Methods for detection of organic contaminants are mostly based on laborious sample processing procedures as a preliminary step to chromatography. In particular the use of liquid-liquid extraction followed by GC-MS chromatography. Speed of analysis is not the prime requirement and sample throughput is often low. While liquid-liquid extraction (LLE) is the favoured preparative method, it suffers from a number of weaknesses:

- it is time consuming for large numbers of samples
- difficult to automate
- requires high purity solvents
- is subject to contamination
- recovery and precision may be poor

Safe disposal of solvents is also problematic and expensive.

Routine analytical methods have been optimised for sensitivity and robustness rather than rapid turnaround time or high sample throughput. Detection limits are based upon long term consumption and to ensure compliance with quality standards rather than meeting short term emergency needs. Nor do current methods recognise the significant advances which have recently been made in chromatography methods.[6] Coupled with this is the fact that LC-MS remains an under-used technique for routine water testing despite advantages in direct analysis of water samples, while derivatisation coupled to GC-MS is still the preferred technique for analysis of low volatility analytes. The reason for the slow take-up of LC-MS is due partly to a lack of standardised equipment in routine water testing laboratories, combined with a general shortage of LC-MS expertise and a lack of reference library spectra. This situation persists despite the maturity of LC-MS as an analytical technique.[7]

At the present time the assessment of microbial quality of drinking water is based almost exclusively on time-consuming culture techniques. Microbial assessment depends upon detection of very low numbers of organisms since many waterborne pathogens exhibit an exceedingly low infectious dose which can pose a hazard to human health. Conventional culture techniques can take many hours or even days to complete.[8] It is also usual for microbial 'indicators' of faecal pollution to be used to monitor for possible presence of human enteric pathogens (since most natural pathogens in drinking water are generally faecal in origin). These

---

* Radioactive contaminants (not considered in this paper) are also covered by recently-developed gross-alpha and gross-beta rapid methods.

indicators do not identify specific hazards, nor do they guarantee to detect the presence of disease-causing vectors. For example, harmful *E.coli* O157 may be present even if faecal coliform determinations give a negative result. A further complication is that many microorganisms are not easily cultured or may occur in a viable but non-culturable state. Alternative methods are therefore required to assess the microbiological quality of water in emergencies.

Indicator culture methods also fail to highlight harmful toxins such as those produced by blue green algae. Facilities for rapid detection of toxins such as microcystins and botulinum are relatively rare within the water companies and methods remain largely in the realm of a few prestigious R&D laboratories.

Existing water quality monitoring for regulatory purposes therefore cannot fulfil an early warning function and this has necessitated the development of new analytical protocols.

3 REQUIREMENTS FOR RAPID ANALYSIS

Speed of analysis is a vital aid to rapid decision making in the case of contamination emergencies. The main priority is to be able to react quickly to prevent or minimise health impact. This means a rapid turnaround time from receipt of samples and requires analysis times ideally of no more than 5-10 minutes per sample. Methods also need to be compatible with delivering a 24/7 response emergency call out capability, ideally with high sample throughput. Chain of custody is considered secondary to (although need not be inconsistent with) speed of analysis. While it is important to be able to detect high priority determinands (established through risk assessment), a broader scope is also desirable (preferably including an ability to detect unknowns).

There are a range of benefits to be gained from employing rapid methods, including the following:

- Early warning of contamination events
- Exercising duty of care to protect consumers & maintain quality
- A deterrent to negligent dischargers and other potential polluters
- Initiating remediation methods
- Identifying sources of contamination
- Process control monitoring and improved operation
- Cost benefits (legal liabilities, protecting works)
- Brand image/avoiding adverse publicity
- Improving business resilience and efficiency

Some individual rapid tests may be suitable for field use. However because of the wide variety of harmful chemical substances and microorganisms that could potentially contaminate a public water supply, there is a need for broad-spectrum rapid tests with good sensitivity (typically 1-10µg/l for chromatographic methods). This requires transport of samples to the laboratory for analysis.

A selection of candidate rapid methods is summarised in Table 1 below.

Table 1. Rapid Methods of Analysis - Summary

Methodology	Chemicals	Toxins	Microbes
Targeted analysis	ICP-MS GC-MS (enhanced) LC-MS (enhanced)	ELISA	Rapid PCR
Screening techniques	Spot test kits Immunoassay UV spectroscopy Raman (SERS) Toxicity assay	Immunoassay	Metabolic fingerprint Coliform tests Bacterial ATP

[dstl]

20 November 2003

© Dstl 2003

Dstl is part of the
Ministry of Defence

4 OPTIMISATION OF CHROMATOGRAPHIC METHODS FOR TARGETED ANALYSIS

A number of approaches can be taken to enhance chromatographic techniques so as to maintain the advantages of targeted analysis to identify individual analytes of concern. These are summarized below and described in more detail in the next chapter.

4.1 More efficient sample processing

While historically, LLE has been the favoured method of sample preparation for chromatographic analysis, there are a range of other procedures available.[9] Solid-phase extraction (SPE) methods have been available commercially for 25 years and offer a range of sorbents and formats (such as cartridges and disks).[10] They are more convenient to use and reduce solvent waste.

Volatiles and semi-volatiles in solution can also be sampled using solid-phase microextraction (SPME) prior to GC analysis.[11,12,13] Like SPE, this is a well established technique and is quick and does not require solvent, although usage can be more problematic than SPE. It can be employed either by direct immersion into the sample or by exposure to the headspace vapour above the sample.

Stir bar sorptive extraction (SBSE) is a more recent innovation in which analytes can be extracted from aqueous solution using a magnetic stir bar coated with a thick film of partitioning phase material.[14] This provides much higher capacity than SPME fibre. While stirring the sample solution, the stir bar efficiently extracts organic components from the sample. After stirring the bar is removed and the absorbed compounds either thermally desorbed for GC analysis or desorbed into a small liquid volume for LC analysis. SBSE devices are commercially available from Gerstal GmbH under the name "Twister"™, including a Twister Automation Option that enables automated analysis of up to 196 Twister stir bars.

4.2 Improvements in chromatography.

4.2.1 Large volume injection

New injection systems are available for GC which accept larger liquid volumes and concentrate these prior to introduction onto the separating column by venting out the solvent under controlled temperatures and gas flow rates.[6,15] This permits injection of volumes of up to $500\mu l$ compared with standard injection systems which typically allow the transfer of less than $5\mu l$, with significant increase in sensitivity. Large volume injection is also compatible with full scan mode MS for improved detection limits and scope.

4.2.2 Fast GC and Fast LC methods

Reduced run times can be achieved by adoption of Fast chromatography techniques. Fast GC-MS can be achieved with smaller diameter capillary columns, reproducible fast temperature programming (up to 300°C/min) and higher gas flow rates (<5 min per run).[16] Fast LC-MS can also be achieved by means of narrower bore columns.[17,18] These can be combined with use of ion pair reagent mobile phase modifiers and ultra-fast solvent gradients plus higher temperature column conditions.[19] This enables analysis time for many separations to be significantly reduced, typically by 60-75%. Also, because analytes spend less time in the column, band spreading is reduced, with improved sensitivity. There is also reduction in solvent waste.

4.3 Automation

Automation can improve speed, accuracy and simplicity of sample handling. Integration of injectors and autosamplers provides fully automated analysis, for consistent results and in order to free up the analysts' time for other tasks while samples are being processed.

Automated SPE can deliver extraction solvent and eliminates multiple transfer steps. This leads to improved reproducibility and sample throughput, as well as simplifying sample handling procedures.   An example of equipment is the Horizon Technologies SPE-DEX® 4790 Automated Solid Phase Extraction System.

Combining high volume injection GC-MS combined with an automated SPE process enables analysis of the entire sample. It is also possible to incorporate intermediate wash steps prior to elution with solvent, and there is the possibility of carrying out in-situ derivatisation prior to chromatography.

SPME can also be readily automated for use with GC. The ATAS Optic 2 injector coupled to a FOCUS autosampler is usable with SPME as well as a range of other techniques, including headspace, large volume injection and stir bar desorption. The injector system is compatible with a range of common GC-MS systems.

SPE can be fully integrated with LC-MS equipment.[20] For example, a complete system for LC-MS online elution might comprise the PROSPEKT in-line SPE module combined with a solvent delivery unit (SDU) and autosampler (e.g. MARATHON XT™ or TRIATHLON™). A further advantage of online SPE is in providing just-in-time processing, thereby minimising sample degradation.

4.4 Modern instrumentation

4.4.1 Benchtop LC-MS.

High performance benchtop LC-MS instruments are available which are easier to use, making them more competitive with GC-MS. They are particularly useful for use by less experienced operators once rapid method protocols have been developed.

4.4.2 Time-of-Flight mass spectrometry (TOFMS)

Additional capability is offered by new TOFMS instrumentation coupled to Flash-GC or Fast-GC. Transfer of chromatographic methods to TOFMS equipment can potentially increase productivity and quality of the mass spectra of contaminants found which, with the accompanying software, are capable of acquiring over 100 mass spectra per second. An example of new equipment is the Leco Pegasus 4D™ GCxGC TOFS. Use of GC-GC/TOFMS, which employs two columns of different polarities in series can be used for much improved separation capability over a single column without compromising speed of analysis.[21]

4.3 Data processing techniques

New data analysis software can undertake spectral searches using library databases, and perform spectral deconvolution and subtraction to assist in the identification of component mass spectra.[22] This facilitates equipment operation by less skilled staff, freeing up specialist staff to concentrate on a more thorough assessment of the preliminary findings. An example of the use of AMDIS deconvolution software for GC-MS with mass spectral library data supplied by NIST is given in the following chapter (this handles files from various instrument manufacturers).

5 RAPID MOLECULAR METHODS FOR IDENTIFICATION OF PATHOGENIC MICROORGANISMS AND TOXINS

Conventional detection methods fail to provide a timely and adequate assessment of the risk from waterborne disease or to accurately reflect water quality and treatment requirements.[9] Recent advances have been made in a range of new molecular detection technologies based on immunological or gene probe methods which are applicable to water testing.[23] In the Polymerase Chain Reaction (PCR), sample DNA is combined with DNA polymerase, nucleotides, and DNA primers that are specific for a given nucleotide sequence. By applying a rapid cycle of heating and cooling it is possible to replicate the DNA sequence and so amplify the signal, creating millions of copies. The amplified DNA can then be detected by a variety of methods. As little as a single copy of a particular sequence can be specifically amplified and detected. The development of real-time quantitative PCR has eliminated the variability traditionally associated with quantitative PCR, allowing the routine and reliable quantitation of PCR products. Fast PCR involves rapid cycling and replication of DNA from specific microorganisms and can also be used for quantitative assay.

Enzyme-Linked Immunosorbent Assay (ELISA) can be used for detecting particulate material but is particularly well suited to detecting toxins in water [24]. In this technique, specific binding of the labeled antibody (or antigen) is detected by reacting an enzyme label with a substrate that generates a visible colour in the reaction mixture and the results read by photometry. ELISA can used to provide a quantitative estimation of the amount of analyte present. Immunoassays are also available in microwell as well as hand held test strip formats. Additional sensitivity for detecting biological material in aqueous media can be obtained using a new immunomagnetic electrochemiluminescence (IM-ECL) based immunoassay technique[25] employed in ORIGEN® equipment developed by IGEN International Inc. Other immunoassay enhancement techniques include use of fluorescence (e.g. RAMP™ from Response Biomedical Corp.) and magnetic assay detection (e.g. Quantum Design Magnetic Assay Reader).

An advantage of using PCR and immunological methods is that, unlike culture techniques, they do not depend upon viability. They will therefore detect microorganisms that are not easily cultured or which can enter a viable but non-culturable state.

6 SCREENING TECHNIQUES

Screening techniques are useful as a first level of testing. However most lack the discriminating power and reliability of targeted methods. A number of possible approaches are reviewed briefly below.

6.1 Chemical spot tests

Individual spot test methods can be useful in tracking selected contaminants such as cyanide or arsenic but the scope of commercially available tests is limited, especially for organic contaminants. Spot tests are best applied where the cause of the pollution problem is already suspected or in follow-up monitoring (the utility of the test being defined by its particular performance capabilities).

6.2   Immunoassays

Tests are commercially available for a number of selected common pesticides and microcystins, including quantitative as well as qualitative tests applicable for field use (e.g. Strategic Diagnostics Inc.). However the range and capabilities of these tests is somewhat limited.

6.3 Rapid culture methods

The duration of culture-based methods can be shortened by detection of specific enzyme activity associated with the target organism. Results are available within 18-24 hours and are therefore faster than standard methods where confirmation of results is required, which take 48-72 hours. These tests are therefore often referred to as rapid methods although they do not strictly meet the timescales required for emergency response. Monitoring of β-galactosidase activity is used to estimate the number of cultural coliforms. Further specificity is provided by application of differential growth temperatures of 44°C for thermotolerant coliforms/*E.coli* and 35°C or 37°C for total coliforms. The Colifast® Analyser is a semi-automated apparatus used to provide early warning indication of faecal contamination.[26] The growth medium contains 4-methylumbelliferyl-β-D-galactosidase substrate which produces a fluorescent product when hydrolysed in the presence of coliform bacteria. Detection is within 2 hours for high levels ($\geq$100 cfu) but up to 11 hours for 1 cfu.

6.4 Metabolic fingerprint

Biochemical characterization of bacteria can be used as an aid to identification. An example of a commercial system is the Microlog Biolog System. This tests the ability of an unknown bacterium to utilise or oxidise compounds from a preselected panel of different carbon sources contained in a 96-well microplate format. Tetrazolium violet is the redox dye used to colorimetrically indicate utilization of the carbon source in each well. This results in a characteristic pattern of purple wells which, according to the manufacturer, enable presumptive identification of up to 339 species of gram-positive and 526 species of gram-negative aerobic bacteria. There is also a "Dangerous Pathogens Database" which includes some bacterial species considered to pose a potential threat from the deliberate contamination of water supplies. The system requires an incubation step (4-6 hours and/or 16-24 hours) to allow respiration to take place in the wells containing carbon sources that can be oxidized and for the cells to reduce the tetrazolium dye to form the purple colour. The pattern of wells is analysed by the Biolog computer software that seeks to match the pattern to a library of bacterial species and if an adequate match is found, an identification of the unknown isolate is performed although the accuracy of the identification cannot be verified. The system may assist in highlighting unusual bacteria during periods of concern although it cannot be classed as a rapid test in the same context as Rapid PCR or ELISA.

6.5 UV Spectroscopy

UV Spectroscopy involves spanning the spectral range of 230-320nm using a diode array spectrometer. Data for samples is compared with a reference spectrum. Many substances exhibit UV absorption in the wavelength range of 230-320nmwhich produce very wide spectra with varying wavelength maxima. The UV spectrum is affected by organic content, suspended solids and matrix effects which has tended to limit the use of UV spectroscopy as a method for monitoring organic compounds in water. It can however be useful as a means of monitoring changes due to transient contamination events and has potential as an online monitoring method for intake protection. The reference sample could be taken "upstream" and compared with a "downstream" sample. Spectral subtraction reduces signal noise and improves detection limit.

Figures 1-8 show UV spectra obtained with an Analytik Jena AG Specord S100 Diode Array Spectrophotometer fitted with 40 mm path length cells (although up to 100mm path length can be used). Data is presented for asulam and paraquat in tap water as examples. Use of spectral subtraction and 1st derivative provides improved detection (compare Figures 5-8 with Figures 1-4). Measurements are also shown for asulam in river water using a Shimadzu Multi-spec 1500 fitted with a 10mm cell (Figures 9-11). This produced a poorer baseline compared to tap water (compare Figures 1 and 10) but much improved using 1st derivative (cf Figure 11 and Figure 2).

The use of UV diode array scanning is a simple rapid screening technique requiring less than 30 seconds per measurement and is capable of detecting to below 0.05mg/l for many UV absorbing pollutants, including paraquat and diquat in treated waters. It therefore offers a rapid method with minimal sample handling and no sample preparation, while capable of achieving trace detection levels. It is applicable to at-site monitoring as part of a monitoring network and could find uses in groundwater monitoring as well as in checking various points in the distribution process.

6.6 Surface Enhanced Raman Spectroscopy (SERS)

A Raman spectrum is similar to an infrared spectrum and provides information about molecular vibrations that can be used for sample identification and quantitation. Practically, it involves focusing a monochromatic light source (laser beam) on the sample and detecting the inelastically scattered Raman radiation via a spectrometer.[27] This leads to a change in vibrational, rotational or electronic energy of the analyte. A detector then converts photon energy into electrical signal intensity. The majority of the scattered light is of the same frequency as the incident radiation. This is known as Rayleigh or elastic scattering. A very

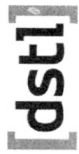

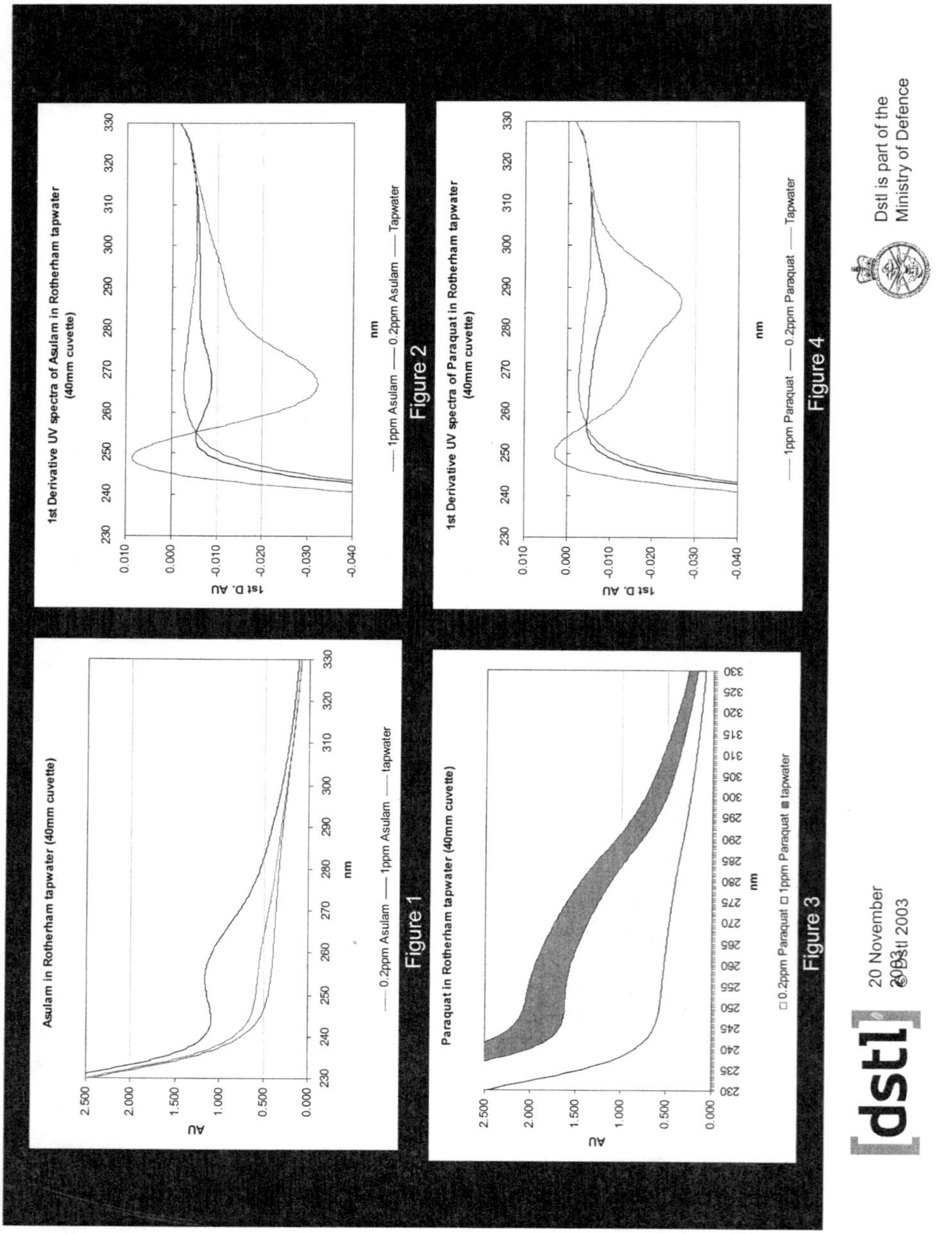
Asulam in Rotherham tapwater (40mm cuvette)
AU
nm
0.2ppm Asulam — 1ppm Asulam — tapwater
Figure 1
1st Derivative UV spectra of Asulam in Rotherham tapwater (40mm cuvette)
1st D. AU
nm
1ppm Asulam — 0.2ppm Asulam — Tapwater
Figure 2
Paraquat in Rotherham tapwater (40mm cuvette)
AU
nm
0.2ppm Paraquat 1ppm Paraquat tapwater
Figure 3
1st Derivative UV spectra of Paraquat in Rotherham tapwater (40mm cuvette)
1st D. AU
nm
1ppm Paraquat — 0.2ppm Paraquat — Tapwater
Figure 4
dstl
20 November 2003
© Dstl 2003
Dstl is part of the Ministry of Defence

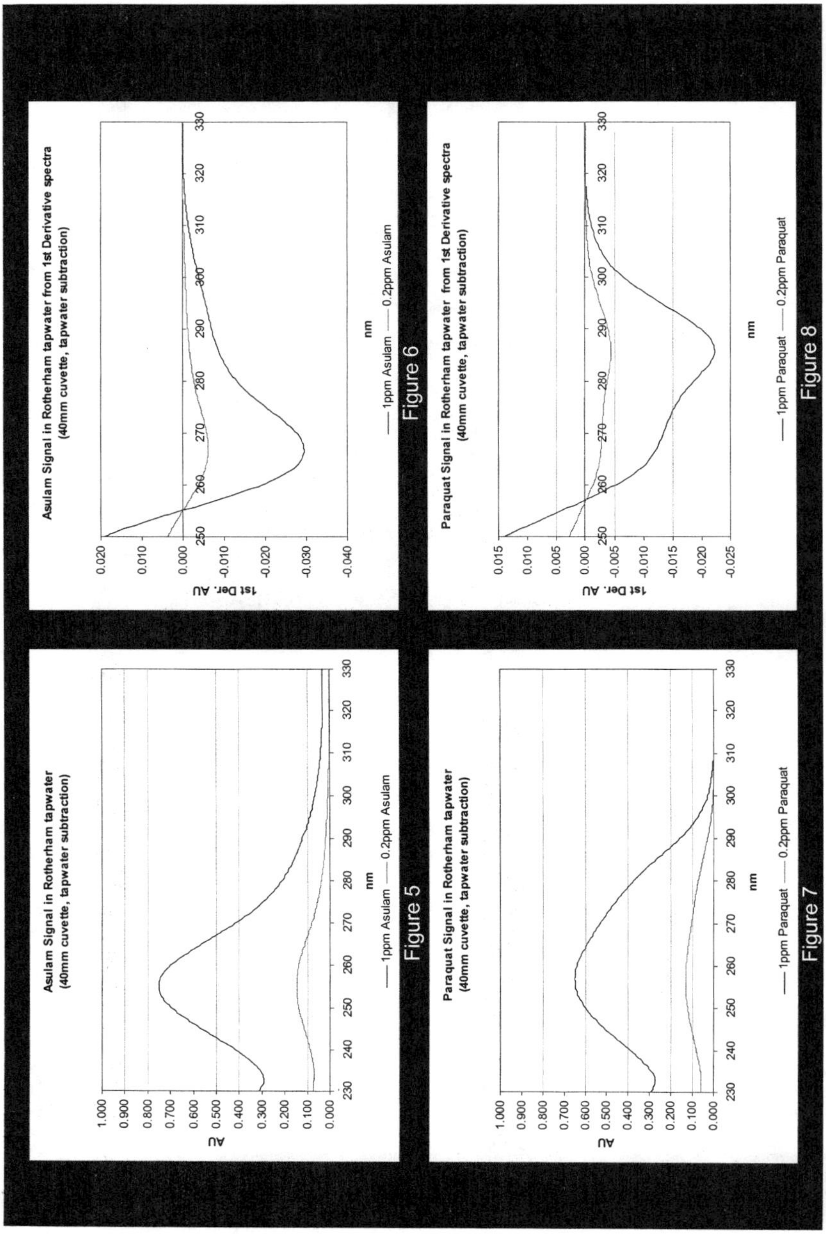
Asulam Signal in Rotherham tapwater
(40mm cuvette, tapwater subtraction)
AU
nm
1ppm Asulam
0.2ppm Asulam
Figure 5
Asulam Signal in Rotherham tapwater from 1st Derivative spectra
(40mm cuvette, tapwater subtraction)
1st Der. AU
nm
1ppm Asulam
0.2ppm Asulam
Figure 6
Paraquat Signal in Rotherham tapwater
(40mm cuvette, tapwater subtraction)
AU
nm
1ppm Paraquat
0.2ppm Paraquat
Figure 7
Paraquat Signal in Rotherham tapwater from 1st Derivative spectra
(40mm cuvette, tapwater subtraction)
1st Der. AU
nm
1ppm Paraquat
0.2ppm Paraquat
Figure 8
[dstl]
20 November 2003
©Dstl 2003
Dstl is part of the Ministry of Defence

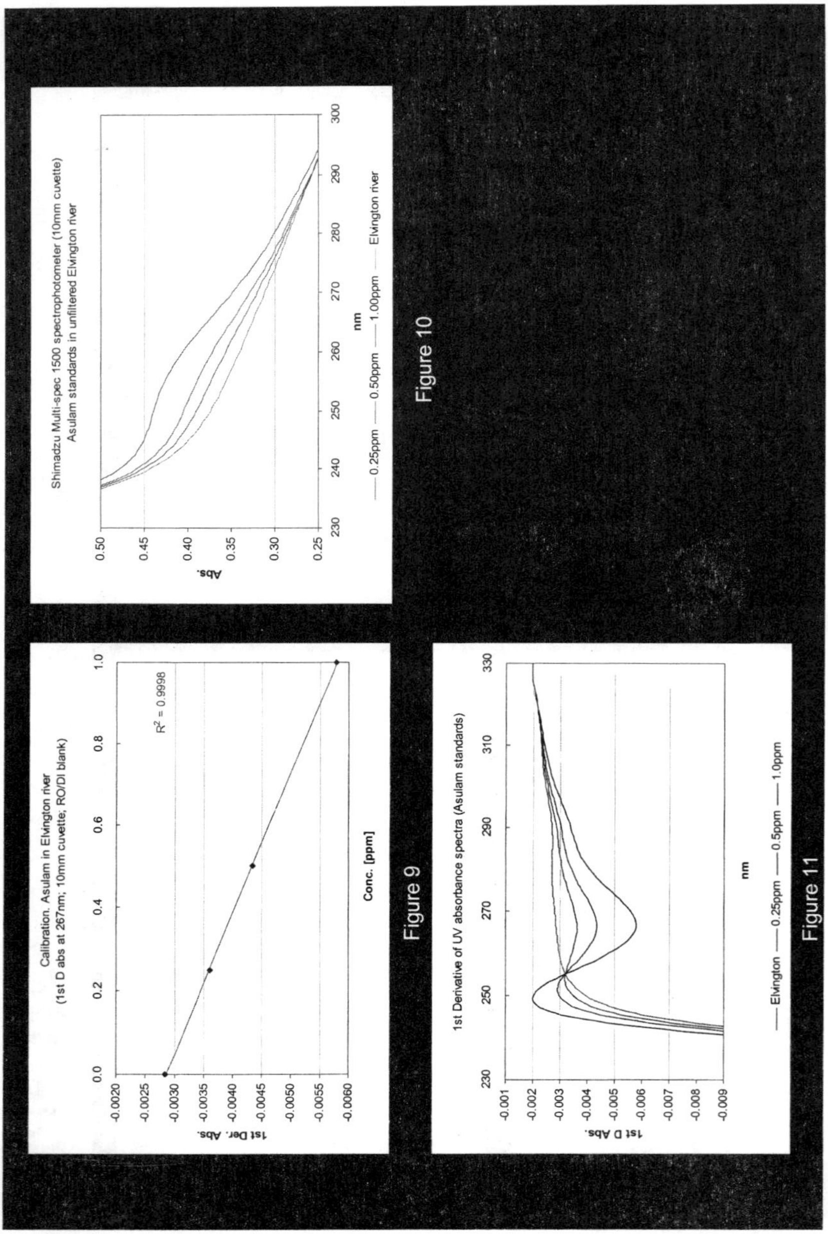
Calibration. Asulam in Elvington river
(1st D abs at 267nm; 10mm cuvette; RO/DI blank)
R² = 0.9998
1st Der. Abs.
Conc. [ppm]
Figure 9
Shimadzu Multi-spec 1500 spectrophotometer (10mm cuvette)
Asulam standards in unfiltered Elvington river
Abs.
nm
0.25ppm
0.50ppm
1.00ppm
Elvington river
Figure 10
1st Derivative of UV absorbance spectra (Asulam standards)
1st D Abs.
nm
Elvington
0.25ppm
0.5ppm
1.0ppm
Figure 11
[dstl]
20 November 2003
©Dstl 2003
Dstl is part of the Ministry of Defence

small amount of the scattered light is shifted in energy due to interaction between the incident electromagnetic waves and the vibrational energy levels of the molecules present in the sample. The fact that normal Raman exhibits a very low conversion of incident radiation to inelastic scattered radiation has limited practical application of ordinary Raman, although it has the advantage over Infrared in providing analysis of bulk aqueous solutions.*

Surface Enhanced Raman Spectroscopy (SERS) is a technique which enables a significant amplification of the Raman scattered light when the analyte is close to, or attached, to a SERS-active surface (typically a coinage metal such as silver or gold). The incident laser photons absorb through surface plasmons within a metal surface, and couple directly with an associated molecule, providing an efficient pathway to transfer energy to the molecular vibration modes, and generate Raman photons. It is a surface technique (100nm) and can enhance the signal by 10^2 to 10^{14} –fold. It has the following advantages:

- Minimal sample preparation is required
- It provides on-demand analysis (suitable for rapid screening)
- Full spectral coverage is generated with every scan
- It has multi-component, quantitative capability
- Spectral subtraction and library searching is facilitated
- Robust portable instrumentation is available, with continuous, real-time and remote measurement possible (via fibre optic probe).

Considerable work has gone into developing SERS-active substrate materials and obtaining reproducible enhancement. Various substrate preparation methods have been tested, including in-situ electrolytic deposition of colloidal silver or gold and formation of metal-doped sol-gels (SG-SERS).[28] The latter are available commercially from Real-Time Analyzers as pre-prepared SG-coated sample vials (Figure 12). SERS provides a simple and rapid method of screening samples, with the potential to develop a library of spectral fingerprints.

Figures 13-15 show Raman spectra for paraquat in water over a range of concentrations. The spectra were acquired within 1-60 seconds using a Renishaw RA200 portable Raman instrument employing a fibre probe head with 785nm excitation and detection in backscatter geometry. The non-SERS spectrum (Figure 13) reveals no useful features over the concentration range 10^{-3}-10^{-7}M. Self-aggregated paraquat on gold colloid (Figure 14) shows no spectral features below 10^{-3}M whereas gold colloid aggregated with KCl continues to show spectral features down to 10^{-10}M (Figure 15).

There are areas of this new technique which still remain to be addressed. The SERS mechanism is not precisely understood and signal intensity depends upon a number of factors. For example in Figure 16 the deviation shown in the relationship between intensity of the paraquat signal at 1642cm^{-1} and concentrations below 10^{-8}M is possibly due to changes at the monolayer coverage point (cf no change recorded in the SERS field enhancement). SERS is susceptible to low adsorption kinetics and competitive adsorption. Problems over reproducibility and re-usability of the substrate have in the past limited practical use but are now being overcome (e.g. via automation of sample preparation). Interpretation of spectra is also more complicated than IR or Raman. However the technique has a number of useful features well suited to screening for contaminants in water.[29]

---

* with an ATR-IR probe measurements can be performed ~1-2μm into water.

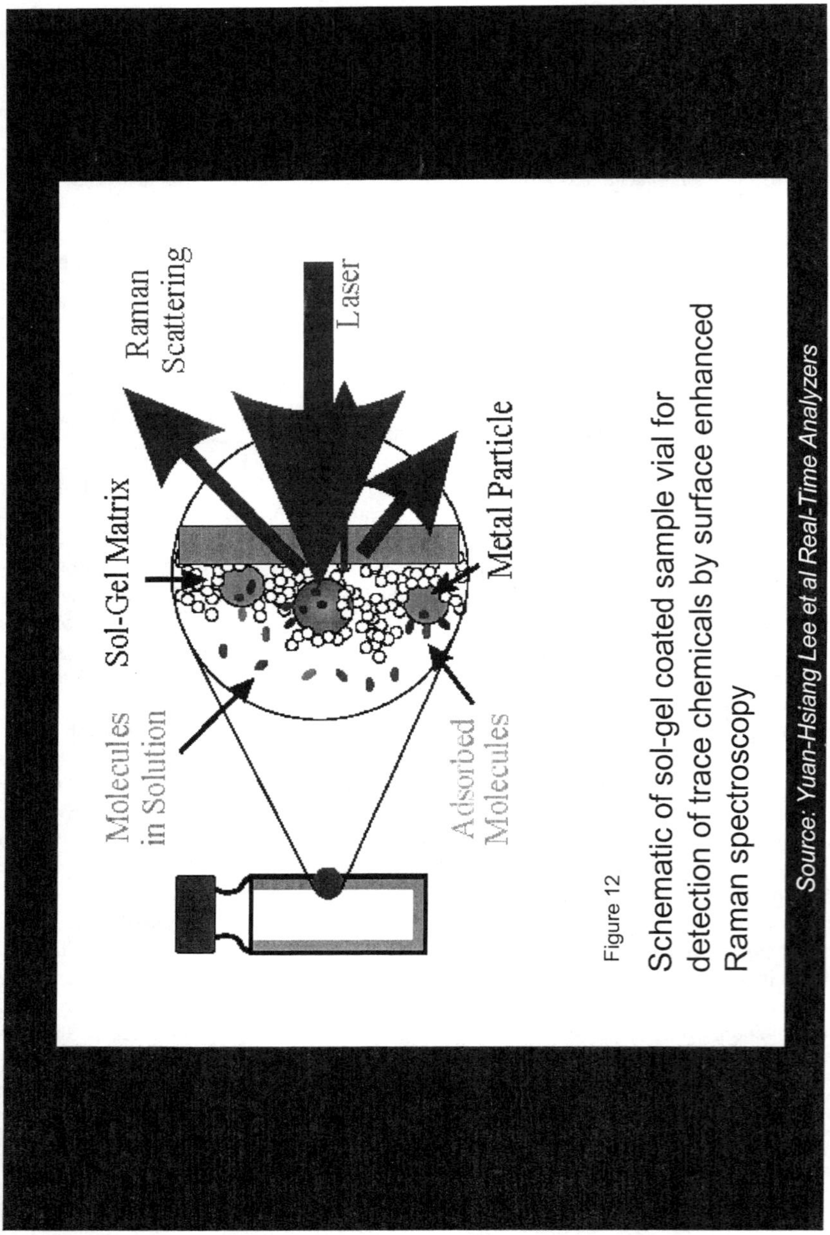

Raman
Scattering

Laser

Sol-Gel Matrix

Metal Particle

Molecules
in Solution

Adsorbed
Molecules

Figure 12

Schematic of sol-gel coated sample vial for
detection of trace chemicals by surface enhanced
Raman spectroscopy

Source: Yuan-Hsiang Lee et al Real-Time Analyzers

Dstl is part of the
Ministry of Defence

20 November
2003   © Dstl 2003

Figure 13

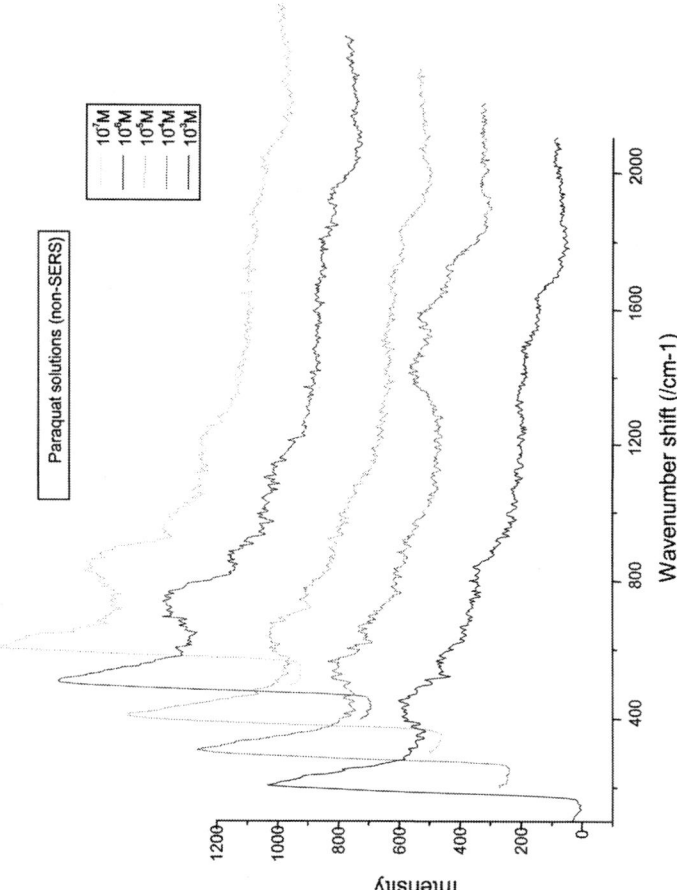

Paraquat solutions (non-SERS)

10⁷M
10⁶M
10⁵M
10⁴M
10³M

Paraquat solution Raman spectra (non-SERS)

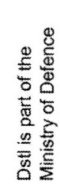

Dstl is part of the
Ministry of Defence

Figure 14

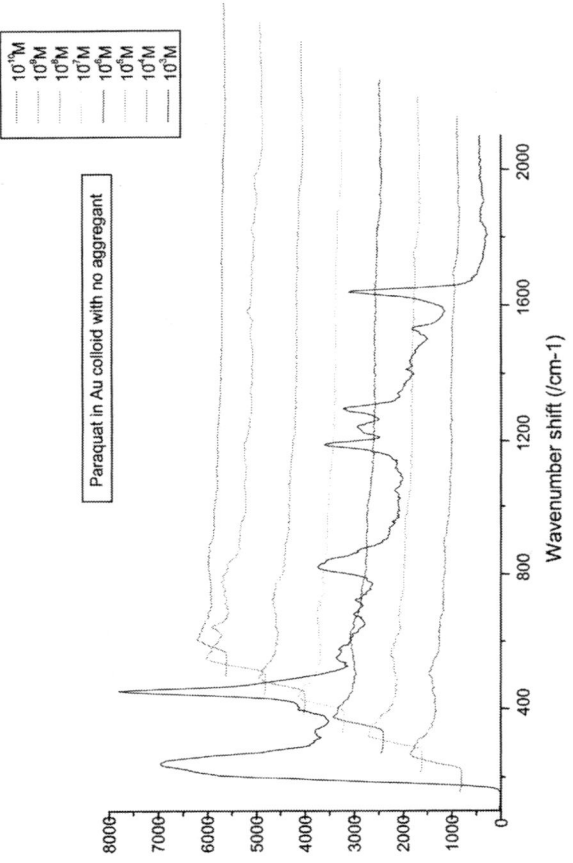

Self-aggregated paraquat on gold colloid SERS spectra

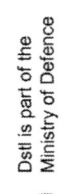

Dstl is part of the
Ministry of Defence

[dstl]

Figure 15

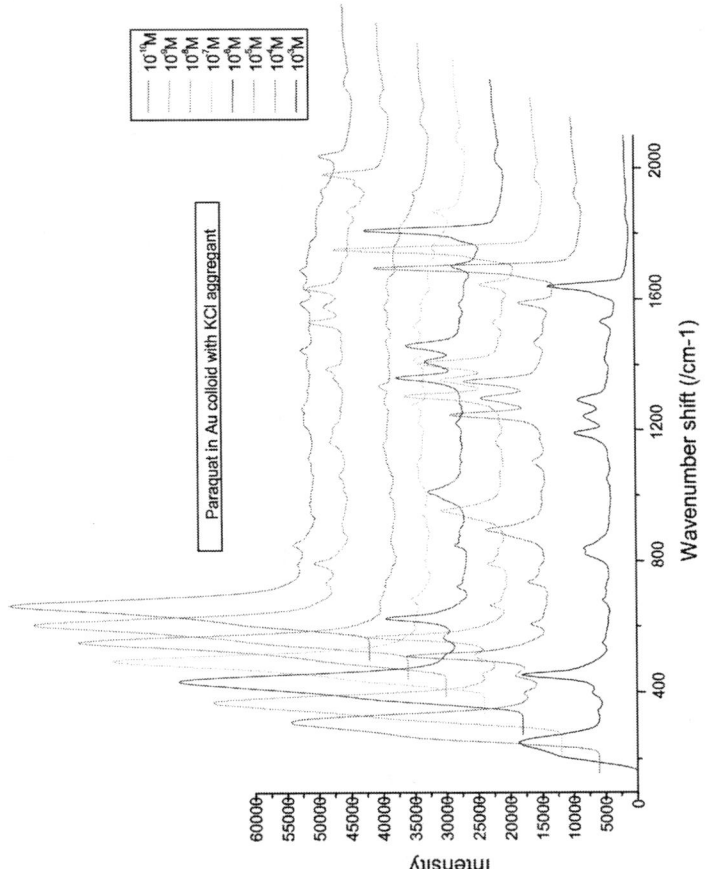

SERS spectra of paraquat on gold colloid aggregated with KCl

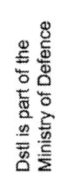

Figure 16

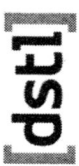

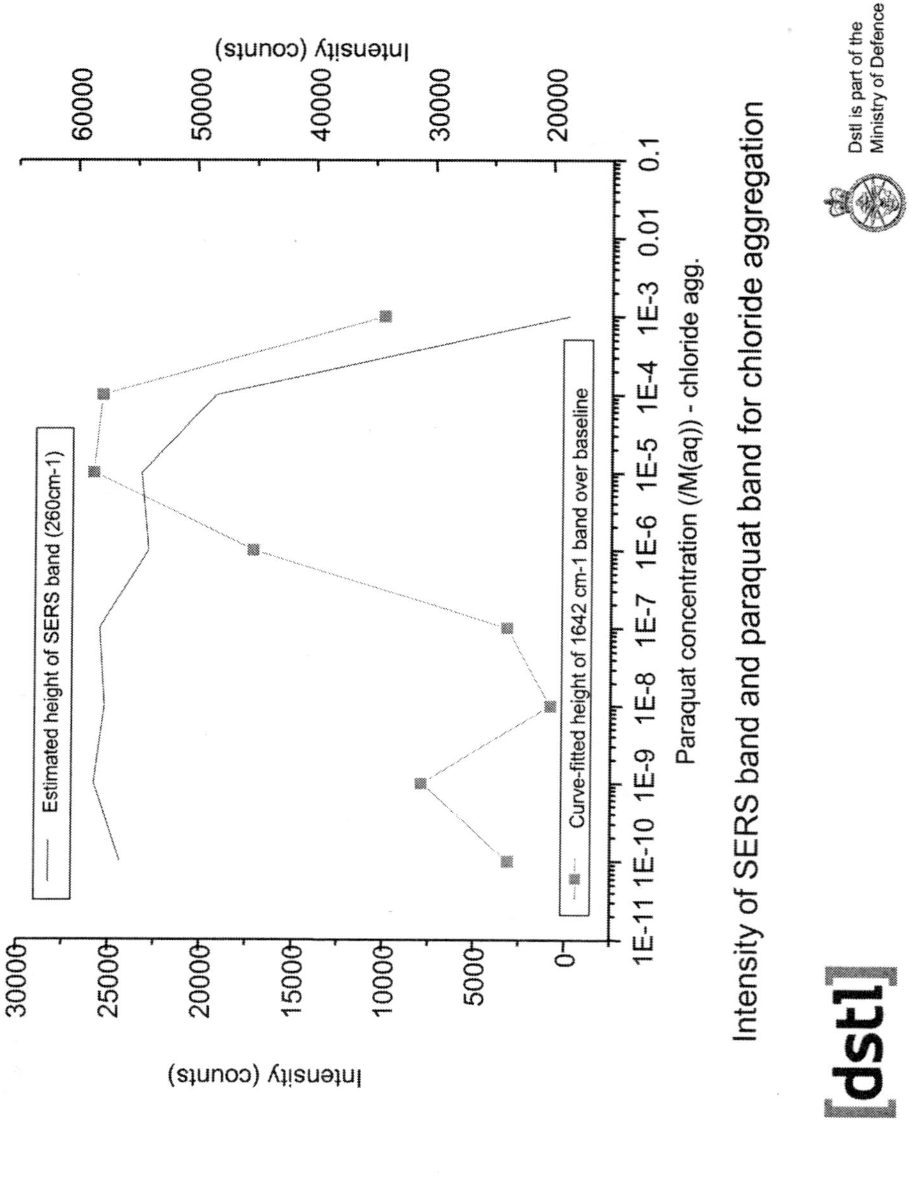
Intensity of SERS band and paraquat band for chloride aggregation
Estimated height of SERS band (260cm-1)
Curve-fitted height of 1642 cm-1 band over baseline
Intensity (counts)
Intensity (counts)
Paraquat concentration (/M(aq)) - chloride agg.
30000
25000
20000
15000
10000
5000
0
60000
50000
40000
30000
20000
1E-11
1E-10
1E-9
1E-8
1E-7
1E-6
1E-5
1E-4
1E-3
0.01
0.1
dstl
Dstl is part of the
Ministry of Defence

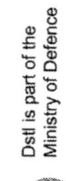

6.7 Other screening techniques

Non-specific indications of a change in water quality can be obtained through use of a range of techniques. These include use of physicochemical parameters such as pH, turbidity, colour, conductivity, as well as chlorine demand, Dissolved Oxygen Concentration, BOD, TOC, headspace (including odour, photo-ionisation detection, or electronic nose sensor array) and toxicity tests. Non-specific indicator sensors are best used on-line in multi-parametric measurements enabling changes to be detected. In-pipe measurements are now possible for a number of parameters (e.g. CENSAR® from Censar Technologies Inc. which enables up to 6 parameters [Electrical conductivity; pH; temperature; redox potential; dissolved oxygen; and free chlorine] to be measured simultaneously).

Various biomonitors have been developed to screen for the presence of toxic substances, sometimes referred to as ecotoxicity tests (Table 2). These range from the well established fish monitor through to enzyme inhibition based sensors, a number of which have already been advocated as general indicators of pollution.[36]

Cholinesterase inhibition tests are available as a non-specific screen for anticholinergic compounds including a range of organophosphorus and carbamate pesticides and are available commercially in different formats, either as quantitative laboratory tests or qualitative field tests (suppliers include EnviroLogix Inc., Neogen Corporation and Machery-Nagel).

Protein phosphatase inhibition can be used to screen for blue-green algal toxins (cyanotoxins), although no commercial test appears to be available.[24,30] Recently it has been claimed that *Thamnocephalus platyrus* (fairy shrimp) bioassay is also sensitive to cyanotoxins.[31] This assay is commercially available as the Thamnotoxkit F™ from Microbiotests Ltd. Along with other Microbiotests available from the same company, it contains the organism in dormant state but needs to be revived in advance in order to be ready for use.

Methods to simplify and automate standard bioassays are being developed commercially and online biomonitors have recently been introduced (examples being the Daphnia-Toximeter™ from bbe Moldaenke and Microtox®-OS from Siemens). However, while toxicity tests can provide a broad screening capability, sensitivity and test time varies according to the contaminant and the test species involved and therefore requires careful selection of the test system(s) used. Running costs can also be quite high and confirmatory testing is still required. The more responsive monitors are the ones best suited to act as online monitors for intake protection.

7 FUTURE DEVELOPMENTS

New analytical developments are taking place, notably in rapid biological detection capability which will simplify test procedures, making them easier to implement, more reliable and provide significant improvements in sample throughput (e.g. through use of microarray technology, DNA gene chip technology and integrated real-time PCR).[32,33,34] Advances will continue to be made in concentration techniques and protein extraction methods leading to improvements in test sensitivity. Chemical detection capability will also extend to monitoring for a wider range of analytes. For example through the use of monolithic LC columns in

Table 2. Examples of biosensors for screening

Specificity	Laboratory	Online	Field
Broad (fish, yeast, bacteria, daphnia, algae, mussels, enzymes)	Microtox® Cellsense Toxkit™ Gentronix test MAAFA	Bio-Sensor® KEMA Mussel Monitor Daphnia-Toximeter Microtox®-OS Amtox™	ToxAlert® Deltatox™ PS1 CheckLight Eclox™ Baroxymeter
Selective (cholinesterase enzyme inhibition)	EnviroLogix Inc ChE screen (microwell plate)		Agri-Screen® pesticide test ticket
Specific (immunoassay)	EnviroLogix assays for pesticides	Biacore SPR	SDI tests pesticides, microcystins Tetracore BT Strips

Dstl is part of the
Ministry of Defence

20 November
2003 © Dstl 2003

ultrafast high-resolution LC/MS/MS.[35] Further improvements are also expected in sample processing and data handling.

Online early warning systems as a trigger for sampling and analysis are likely to improve and be more widely used in future for intake protection, enabling changes in water quality to be detected indicative of a contamination event. Many of the latest advances in sensors, especially those applied to biological detection, are derived from recent impetus in defence research and the need to ensure security of public water supplies.[36]

8 SCREENING VS TARGETED ANALYSIS

Screening and rapid target analysis both have a role to play in water contamination emergencies. Targeted analysis requires development of rapid methods optimised for efficient sample throughput, while screening tests can be employed in the reception laboratory to identify and prioritise samples prior to targeted analysis. When processing large numbers of samples, both techniques may be operated in parallel for maximum efficiency. Targeted methods for priority chemical contaminants can be based on Fast GC-MS and LC-MS, while Rapid LightCycler-based PCR is suited to detection of pathogens and ELISA to toxins. Techniques of current interest for rapid pre-screening include UV and SERS spectroscopy. At the present time there are few comprehensive online monitoring systems or rapid, broad-spectrum screening procedures available for assessing the potability of drinking water. Hence the reliance on laboratory based analysis.

9 EMERGENCY CALL OUT CAPABILITY

Testing capabilities must be based on an appropriate risk assessment, while rapid analytical methods need robust, reproducible, clearly-defined protocols. Once methods have been established and validated, they need to be well-rehearsed and preferably in regular use. This will ensure a high state of readiness and provide for staff training requirements. Equipment should also be well-maintained and in regular use with spares and backup capability. Triggering procedures for the call out should be clearly understood, and sampling requirements established in advance. Routine QC performance testing should also be carried out to maintain confidence. These procedures should be backed up by occasional exercises to test the system, particularly when call-out events are infrequent. This will also provide a check on communication procedures and instructions to staff.

Maintaining a 24 hour, year-round on-call capability and reception facility is an expensive investment for an individual water company, requiring access to a large pool of experienced analysts (including chemists, microbiologists, toxin immunoassay analysts, and pcr analysts). Reagents, standards and analytical equipment must be maintained in a state of readiness. A turnover of limited shelf life reagents will need to be accommodated, combined with ready access to new stocks of consumables. It may therefore be appropriate to contract the task out to specialist laboratories familiar with the handling of toxic materials, which possess the facilities and capacity to respond to an emergency at any time of day or night.

Acknowledgements

Thanks go to Dr Chris Dyer (Dstl) for providing the SERS examples and Professor Clive Thompson (ALcontrol Ltd) for supplying the UV spectra.

The opinions expressed in the paper are the author's own and references to commercial equipment is included by way of example and is not intended to provide an endorsement of any manufacturer's product.

References

1   Water Supply (Water Quality) Regulations 2000. The Stationery Office Ltd. ISBN: 0 11 018878 0.
2   Methods for the Examination of Water and Associated Methods. Series published by the Standing Committee of Analysts. (listed at http://www.dwi.gov.uk/regs/pdf/scabb202.pdf).
3   The Microbiology of Drinking Water (2002) – Parts 1-10. Methods for the Examination of Waters and Associated Materials. Environment Agency.
4   J.G. Holland and S.D. Tanner. Plasma Source Mass Spectrometry: Applications and Emerging Technologies. Royal Society of Chemistry (2003) ISBN: 0 85404 603 8.
5   J.S. Becker, *Can J. Analytical Sciences & Microscopy*, 2002, **47**(4): 98-108.
6   R.B. Geerdink, W.M.A. Niessen and U.A.T. Brinkman, *J Chromatog. A*, 2002, **970**(1-2): 65-93.
7   E. Hogendoorn, and P. van Zoonen, *J Chromatography A*, 2000, **892** (1-2): 435-453.
8   Rapid Microbiological Methods: The Status Quo. The Blue Pages. International Water Association. July 2000. (www.iwahq.org.uk).
9   G.A. Mills, *Chromatography and Separation Technology*, 2003, Issue 26 pp18-21 (Jan/Feb).
10 M.E.C. Queiroz, S.M. Silva, D. Carvalho et al, *J. Environ. Sci. Health B-Pestic*, 2001, **36** (5): 517-527 2001.
11 B. Zygmunt, A. Jastrzebska and J. Namiesnik, *Critical Rev. Anal. Chem.*, 2001, **31**, (1) 1-18, Review.
12 A.A. Boyd-Boland, S. Magdic and J. Pawliszyn, *Analyst*, 1996, **121**, 929-938.
13 C. Goncalves and M.F. Alpendurada, *J Chromatography A*, 2002, **968**(1-2), 177-190.
14 E. Baltussen, P. Sandra, F. David and C. Cramers, *J Microcolumn Separations*, 1999, **11**(10): 737-747.
15 F.J. Santos and M.T. Galceran, *J Chromatog A*, 2003, **1000**, (1-2): 125-151.
16 K. Mastovska and S.J. Lehotay, *J. Chromatog. A*, 2003, **1000**, (1-2): 153-180
17 T. Wehr, *LC-GC North America*, 2002, **20**(1) 40-47. See also Dong M.
18 The time is now for fast LC. Today's Chemist at Work. American Chemical Society p46-51 (February 2000).
19 T. Greibrokk and T. Andersen, *J. Chromatography A*, 2003, **1000**(1-2): 743-755.
20 T. Koal, A. Asperger, J. Efer and W. Engewald, *Chromatographia*, 2003, **57**, S93-S101.
21 C. Leonard and R. Sacks, *Analytical Chemistry*, 1999, **71**(22): 5177-5184.
22 A. Robbat, S. Smarason and Y. Gankin, *Field Analytical Chemistry and Technology*, 1999, **3**(1): 55-66.

23 N. Lightfoot, M. Pearce, B. Place and C. Salgado in Rapid Detection Assays for Food and Water. Royal Society of Chemistry Special Publications (2001), 59-65.

24 J. Rapala, K. Erkomaa, J. Kukkonen, K. Sivonen and K. Lahti, *Anal. Chim. Acta*, 2002, **466**(2): 213-231.

25 E. Kuczynska, D.G. Boyer and D.R. Shelton, *J. Microbiological Methods*, 2003, **53**(1): 17-26.

26 I. Tryland, S. Surman and J.D. Berg, *Water Science & Technology*, 2002, **46**(3), 25-31.

27 M. Pelletier, Analytical Applications of Raman Spectroscopy. Blackwell Science (1999). ISBN: 0 6320 53054.

28 Y-H. Lee and S. Farquharson. Proceedings of the Society for Photo-optical Instrumentation Engineers (SPIE), 2001, 4206, 140-146.

29 N. Weissenbacher, B. Lendl, J. Frank, H.D. Wanzenbock and R. Kellner, *Analyst*, 1998, **123**(5): 1057-1060.

30 B. Nicholson, *Water Quality News*, 1998, Issue 7, December.

31 B.C. Nicholson and M.D. Burch, Evaluation of analytical methods for detection and quantification of cyanotoxins in relation to Australian drinking water guidelines. NHMRC publication (October 2001).

32 T.M. Straub and D.P. Chandler, *J Microbiological Methods*, 2003, **53**(2): 185-197.

33 T.M. Scott and J.B. Rose, *Water Conditioning & Purification Magazine*, June 2002.

34 K.S. Betts, *Environmental Science & Technology*, 1999, **33**(15): 300A-301A.

35 B. Whitehead, S. Brombacher and D.A. Volmer, Proceedings of the 50th ASMS Conference on Mass Spectrometry and Allied Topics, Orlando, Florida, June 2-6, 2002.

36 S. States, M. Scheuring, J. Kuchta, J. Newberry and L. Casson, *Journal AWWA*, 2003, **95**(4)103-115.

FIELD AND LABORATORY ANALYSIS FOR DETECTION OF UNKNOWN DELIBERATELY RELEASED CONTAMINANTS

G. O'Neill[1], C. Ridsdale[1], K. Clive. Thompson[2] and K. Wadhia[2]

[1]Yorkshire Water Services, Bradford, Yorkshire, UK
[2]ALcontrol Laboratories, Rotherham, Yorkshire, S60 1BZ, UK

1 INTRODUCTION

There is no doubt that the best defence against any deliberate release of chemical, biological or radioactive agents lies in the form of physical barriers and entry alarm systems. In the event that these barriers are inadequate or are somehow breached, then analysis for an unknown potential contaminant becomes critical.

In terms of water supplies, the most obvious point of deliberate contamination is the storage reservoirs, but possibly the most vulnerable point of interference is after the treatment process. An obvious target is the service reservoir. Vandalism at service reservoir sites has always been a regular occurrence but has been elevated to greater significance since September 2001 (nine-eleven). When circumstances give rise to suspicion that contamination may have occurred, samples are taken immediately the awareness of the situation becomes known. Samples are sent to the laboratories, where, at any time of the day or night a range of sophisticated analyses are performed. This service is as good as it can reasonably be and is being improved all the time. Some results are available within an hour of the sample arriving at the laboratory, but to complete the full suite of analyses it takes 16 hours in addition to the travel time. Historically, deliberate contamination has never to date been detected but the question does arise as to how effective our procedure is in protecting our customers against potential deliberate contamination. One possible procedural improvement that could be instigated is to take immediate action in the network on every occasion on the assumption that it is incident of actual and not potential contamination. In many cases this would be hugely expensive and disruptive both to normal operational activity and to the customer.

A possible solution is to have the capability to rapidly detect a wide range of toxicants "at-site", thus eliminating transport time. Recent developments in biotechnology have provided some broadscreen capabilities for this. A system by Capital Controls currently in production, the Eclox-M™, employs chemiluminescence. This has been evaluated for military use in the field, and is already in widespread use. An alternative bioluminescence system, DeltaTox™ is also available as a field kit; this is a development of the laboratory-based Microtox® system. These systems provide a broad screen analysis for toxicity. A range of other technologies are also available, some broadscreen and some specific. This includes radioactivity monitoring, UV spectrophotometry for a wide range of organic compounds (with aromatic rings) and photo-ionisation detection for a broad range of volatile organic substances. In addition there are test kits for toxic pesticides (e.g. cholinesterase inhibition) and specific tests for arsenic and cyanide. A system has also been tested which detects ATP to evaluate any gross microbiological contamination.

2 OBJECTIVES

The objective of this project was to evaluate the technologies for their applicability and relevance to detect contamination of potable water supplies, to determine the ease of use, assess the skill level required to operate each technology, evaluate the accuracy and reproducibility of the systems, and consider implications on cost of ownership.

The main emphasis was on the Eclox-M™, but a limited evaluation was made of the DeltaTox™ together with a wide-ranging review of any other available technologies. Some comparisons were also made with current laboratory systems used for identification of unknown contaminants.

3 TOXICANTS

The starting point for test toxicants can be found in a Department of Health document[1]. Additionally, information on toxicity is available from a wide range of sources, e.g. UKWIR Toxicity Datasheets and websites for IRIS, ExToxNet and Chemfinder. None of these is comprehensive and contradictions between websites are not difficult to find. The number of (malicious addition) toxicants that may be relevant to water is not infinite, perhaps in the region of 200. Important factors are the availability of large quantities that would be necessary to significantly contaminate water supply and the solubility of the toxicants in water. The toxicants can be classified into categories given in Table 1.

Table 1 *Toxicants Relevant to Water*

• Metals
• Rodenticides
• Radionuclides
• Pesticides
• Industrial intermediates
• Natural poisons e.g. ricin, botulinum toxin

The rodenticides are listed as a separate group as toxicity to rodents is often used in assessment of general toxicity, and hence ranks highly although some are also highly toxic to man. However, some are not particularly soluble limiting their danger in the context of water supply.

Some radionuclides in sufficient quantities could present a threat, although radioactive isotopes such as uranium and plutonium are arguably more toxic simply as metals.

Some pesticides with unrestricted use are both highly toxic and water soluble.

There is a wide range of industrial intermediates which may be carried in tankers around the country from which an accidental release is at least a theoretical risk. However, not many would be regarded as highly toxic and are commonly highly odorous which at least makes their presence apparent.

Natural poisons such as ricin and botulinum toxin can in theory be crudely produced with limited technology, but production of quantities to be of significance in terms of contamination of water supply is unlikely. However, botulinum toxin in particular has been produced in large quantities as a weapon, and if, as has been reported widely in the press; thousands of litres are unaccounted for in Iraq, its presence in quantity one or two litre around the world in the hands of terrorists could represent a significant potential hazard in water supply.

4 MICROBIOLOGICAL AGENTS

Very large quantities of vegetative organisms e.g. E.coli O157 would be needed to have a significant impact. Spores, particularly anthrax, could be a problem not only if produced on a large scale as a weapon but also when produced on a 'kitchen' scale basis if the organism was initially utilised for culturing purpose.

5 DELIBERATE RELEASE DEFENCES

The most obvious points for a deliberate release of a toxicant are raw water storage reservoirs. In Yorkshire in particular these tend to be relatively remote and not only is security limited but also recreational access to the environs at some reservoirs is actively encouraged. The first defence is dilution.

Addition of quantities of any toxicant that could be carried or loaded in the boot of a car is likely to be diluted to non-toxic levels by the time it arrives at the treatment works. As an added factor the typical multi-stage treatment will also remove contaminants, chlorine would also be a major factor for microbiological contaminants.

Deliberate contamination of treated water undoubtedly presents a greater threat to health. Service reservoirs are the most obvious point of access and conventional security such as highly secure access lids and alarms are the major defence here. However, even in the event of failure of conventional security dilution can make a major contribution, chlorine will also contribute in the case of microrganisms. There are no readily available contaminants that would be lethal at the concentration that would result in a typical service reservoir, following addition of a quantity that could be reasonably carried by an individual.

Probably the final defence against contamination affecting public health is making the drinking water unavailable to the customer. This would require isolating a potentially contaminated service reservoir and providing an alternative supply ideally through re-zoning, or through tankering, bowsers or provision of bottled water but there are obvious limitations to this approach.

6 RESULTS

A number of different field test kits (listed in Table 2) were evaluated

Table 2 *The Field Test Kits Evaluated*

•	Eclox-M™
•	DeltaTox™
•	Pesticide kits
•	Arsenic
•	Cyanide
•	Radioactivity Monitor
•	Scanning UV Spectrophotometer
•	Photo-Ionisation Detector
•	Turbidity Meter
•	pH Meter
•	Conductivity Meter
•	Biotrace™

6.1   Eclox-M™

The principle of the Eclox-M™ is inhibition of chemiluminescence. Light is generated in the Eclox-M™ system by the reaction of an enzyme (horseradish peroxidase), a chemiluminescent molecule, an oxidant and an enhancer. The light emitted from the reaction is measured by a simple photomultiplier.

The Eclox-M™ system was found to be easy to use and gave a result within 15 minutes as claimed by the manufacturer.

A wide range of toxic substances have been reported as having an effect[2], although amongst the substances used in this evaluation only cyanide was detectable at low concentrations. A concentration of 0.01 mg l^{-1} cyanide resulted in 23% inhibition (Figure 1). However the Galgo Chemets Test that requires no reagent addition was much simpler and more importantly specific. It is considerably more robust with an immediate visual response.

Figure 1 *Inhibition of chemiluminescence by KCN.*

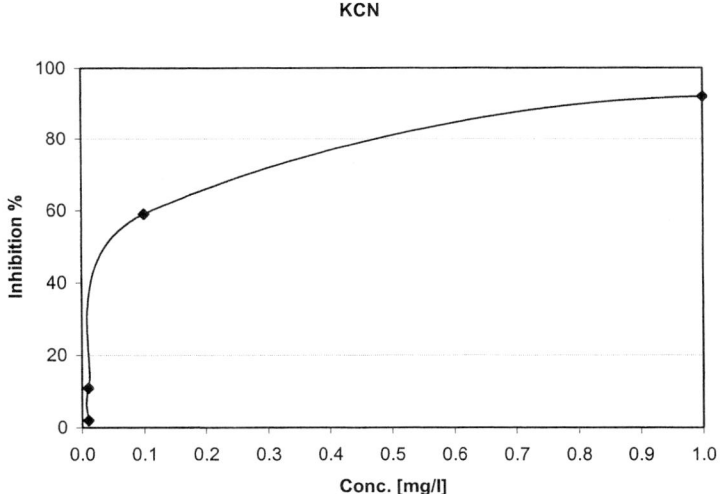

Inhibition by a range of other toxicants can be demonstrated but the concentrations needed are very high for any practical purposes. For example, 300 mg l^{-1} paraquat gave only 22% inhibition (Figure 2), thereby making it of limited use for this particular substance.

Figure 2 *Inhibition of chemiluminescence by paraquat*

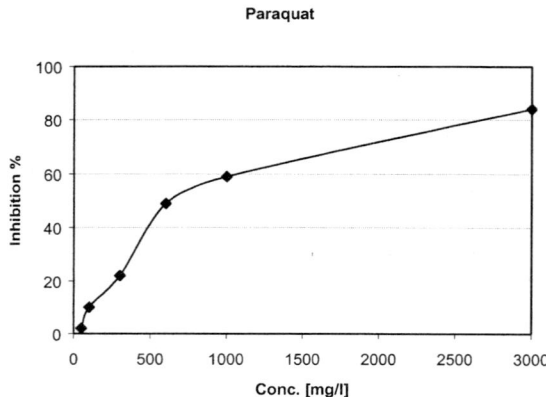

Any naturally occurring constituent component of the test sample likely to mask or interfere with the chemiluminescence reaction will potentially contribute to erroneous results. Manganese is commonly found in natural waters and in order to determine the inhibitory effect of the element, a range of concentrations were tested. Inhibition values ranged from 18% to 65% with a concentration series from 0.06 to 0.25 mg l^{-1} (Figure 3).

Figure 3 *Inhibition of chemiluminescence by manganese*

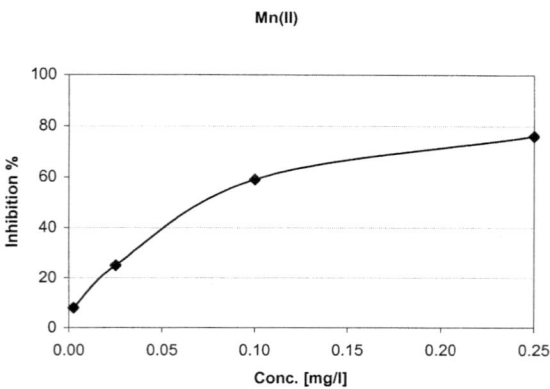

Some additional results are given in Tables 3 and 4. Other substances found to have little effect with the Eclox-M™ were strychnine, nicotine, mercury, a range of solvents, and a range of pesticides.

6.2   DeltaTox™

The DeltaTox™ system is based on bioluminescence of the bacterium *Vibrio fischeri*. In the presence of oxygen and an energy source, the luciferase enzyme contained within the bacterium oxidizes a substrate and one of the end products is light. The DeltaTox™ system is a rapid portable test method that measures the change in bioluminescence.

Like the Eclox-M™, the DeltaTox™ was found to be easy to use and gave a rapid result. However, it is not sensitive to most of the toxicants thought to be important. It detects arsenic at low concentrations and a huge range of substances with toxic effects has been evaluated[3] but little effect could be seen from a range of substance tested in this project.

6.3   Pesticide Kits

Neither of the luminescence technologies detects paraquat. Additionally neither technique was sensitive to a range of pesticides and other toxic substances generally available at outlets such as garden centres. There is an apparent awareness of these limitations by the manufacturers of the Eclox-M™ and the kit includes a test kit for pesticides. This is based on cholinesterase inhibition, essentially the toxic effect of organophosphates and carbamates. The test is reasonably sensitive to these pesticides particularly when used with a bromine activator. Recently a more sensitive technology from Macherey Nagel, also using the principle of cholinesterase inhibition has become available and this is under evaluation.

6.4   Arsenic and Cyanide

Arsenic and cyanide are not highly toxic in terms of the quantities that may be needed to have a significant effect but are worthy of note as well known substances in general sense. Specific kits for both were evaluated, the arsenic kit specifically included with the Eclox-M™ and both were found to be very effective.

6.5   Radioactivity Monitor

The radioactivity monitor, Berthold LB123, can detect gross alpha beta and gamma using two different heads, but detection of alpha particles from uranium or plutonium, the most likely to be used in deliberate contamination, would require very large quantities in water. However, the system is sensitive to low levels in powder form and a rapid means of evaporating water is under development as a field test.

6.6   Scanning UV Spectrophotometer

A scanning spectrophotometer, Shimadzu Multispec-1501, needed very little development work as it is a well-established system that can detect a wide range of organics (containing aromatic ring(s)) at concentrations around $0.1 - 1$ mg/l.

6.7   Photo-Ionisation Detector

The photo-ionisation detector, Ppb RAE, used was not designed for water but for atmospheric detection of volatile substances, including diesel oil. However, a technique has been developed for use with water. Diesel can be detected down to 2ug/l. During the course of the work, an incident occurred in the company involving accidental contamination of water with light mineral oil consisting of aromatic (3%), aliphatic (42%) and naphthenic (55%)

compounds. The PID was used in this incident and found to be highly effective, easy to use in the field and more sensitive than the available laboratory techniques of either IR spectroscopy or even GC – MS.

6.8   Biotrace™

The Biotrace™ system again uses bioluminescence as an indicator dependent on the presence of ATP. Potentially this could be used to detect sewage contamination or even possibly a deliberate release of bacteria. An initial evaluation suggests that it would be on the operating limits of detection and is somewhat inconsistent; it is also not easy to use.

7   EMERGENCY LABORATORY ANALYSIS

Comparisons have been made as part of the project with a laboratory-based broadscreen Toxkit from MicroBioTests Inc, which uses *Thamnocephalus platyurus* (a freshwater crustacean). This picks up wide range of toxicants at relatively low concentrations, but requires 24 hours for tests to be completed[4]. However, effects can be picked up more rapidly with higher concentrations of toxicant and work is in progress to enhance this effect (See chapter ?).

Outside of this project, but being developed simultaneously, is a suite of analyses for laboratory use in situations requiring emergency analysis. This includes broad screen non-specific monitoring including *Thamnocephalus platyurus* and a UV scan, as well as specific analysis for metals and organics, however, even this is not comprehensive. A rapid test for sodium fluoroacetate based on ion chromatography has also been developed. A comparison of the sensitivities from the different technologies is given in Tables 3 and 4.

Table 3 *Comparison of the Sensitivities of the Eclox™, Microtox® and Thamnotox F™ tests (all results in mg/l)*

Chemical	Eclox™ (IC_{50})	Microtox® (EC_{50})	Thamnotoxkit (LC_{50})
Arsenic	26	0.3	1.3
Cyanide	0.07	3.5	0.3
Sodium fluoracetate	>1000	507	0.4
Dichlorvos	354	29	19
Thallium	113	450	0.2
Paraquat	73	20	0.6

Table 4 *Comparison of the Sensitivities of the Eclox™, Microtox®, DeltaTox™ and Thamnotoxkit F™ tests using Proprietary Products (all results in mg/l)*

Product	Eclox™ (IC_{50})	Microtox® (EC_{50})	DeltaTox™ (Acute) (EC_{50})	Thamnotoxkit F™ (LC_{50})
Jeyes Fluid	0.62	0.9	0.49	21
Weed killer	285	422	489	3200
Cuprinol	20262	9.5	3.7	0.3
Rose Clear	2863	1.4	2.3	30
Sheep Dip	20073	1473	688	3200

The concentration of the active ingredient(s) of the proprietary products, is as follows:

–Disinfectant – 'Jeyes Fluid'   •Tar Acids (21% W/W)and Methanol
–Herbicide – 'Lawn Weedkiller' • 2,4-D (38.8g/l) and Dichlorprop (97.0 g/l)
–Fungicide - 'Cuprinol' •Benzalkonium chloride (3.0% W/W)
–Systemic Insecticide & Fungicide -'Rose Clear' •Pirimicarb (22 g/l), Bupirimate (27.8 g/l), Triforine (27.8 g/l)
–'Sheep Dip' – Cypermethrin (1.25% W/W) cis-trans 80/20

8 PRACTICAL CONSIDERATIONS

Each of the technologies listed in Table 2 has a value in assessment of potential unknown contamination. However, it is apparent that limitations arise in any attempt requiring transportation and utilisation in the field. One possible solution is the use of any company building within close proximity of a potential problem site that has a stretch of bench space and a sink.

To make the process practicable a vehicle adapted for use with the various technologies becomes essential and this is being developed.

The advantages of conventional sampling and delivery to the laboratory as an alternative or perhaps in addition will always need to be considered.

9 DISCUSSION

It is apparent from the work carried out that the Eclox-M™ system is limited in its range of toxicants that might be detected. Furthermore, the inhibition by manganese could be the source of serious embarrassment given the high levels of manganese in raw waters, it can be deposited in the network, and even in some sample lines. Hence, the technology will be incorporated in any emergency analysis but results are unlikely to promote any significant response in isolation.

The principles of the DeltaTox™ as embodied in the laboratory based Microtox® have been established for some time and this project has shown that results from the former appear very similar to the latter. But as with the Eclox-M™ there are limitations in its capability to detect toxicants.

The test kits for pesticides are particularly valuable in that although they would not detect all pesticides, the principle of cholinesterase inhibition should detect many of public health significance. The results with the tests included in the Eclox-M™ kit are good but a better technology may be available using a photometer-based system, this will depend on the use of the mobile laboratory.

The arsenic and cyanide test kits are easy to use with a good low-level detection limit. Although neither toxicant is likely to pose a significant threat to public health from water, either could easily be the basis of a hoax threat. Rapid negative results would provide reassurance.

The radioactivity monitor appears very sensitive and is important for rapid analysis for a hoax or serious terrorist threat as radioactive material could easily be in the hands of terrorists. Work is necessary to provide a process needed to remove water from the sample in the field. With the mobile laboratory, this becomes a real possibility.

The one technology that has actually been used in earnest during this evaluation project has been the Photo-Ionisation Detector. This detects volatile organics, is very sensitive to diesel oil (a relatively common source of contamination) and has been adapted for rapid site use with water. It proved to be very valuable following a mineral oil contamination of a water main, not only because of the immediate availability of results but also because of its increased sensitivity over the existing laboratory methods.

Little consideration has been given to the evaluation of pH, conductivity, and turbidimeters but these technologies are readily available and have been used on a regular basis. The availability of the mobile laboratory will enhance their practicability.

The Biotrace system, essentially ATP measurement has been around in various forms for some years. Limitations were encountered since the numbers of bacteria required to be detected are always on the edge of the detection limit of the system and very minor variation in technique is capable of apparently significant variation in result. However, it still may be the case that this technology is worth further evaluation.

Unquestionably the one single factor making the field testing practicable is the use of a vehicle with a bench, sink and availability of a 240V supply, this allows the wide range of field kits to be transported and used on-site. No one of these kits could be relied upon to give a definitive answer and indeed even in their entirety could never be used with 100% certainty. The laboratory will give a much wider range but even that would not cover all possible contingencies.

10 FURTHER PLANNED WORK

Development work is continuing in various areas, including the evaluation of a new system for use as a broad band pesticide screen, field testing of all the systems using the specially adapted vehicle, development of a field method for evaporation of liquids for radioactivity monitoring, and testing of selected toxicants which so far have proved difficult to obtain. It is also proposed to undertake further work with the Biotrace™ system and to develop an enhanced more rapid Thamnocephalus toxicity test.

References

1   Anon. 2000. Deliberate Release of Biological and Chemical Agents. Guidance to help plan the health service response. Department of Health, London.
2   I. Johnson and I. Harman, Environment Agency R&D Technical Report E28, 2002, Environment Agency, Bristol.
3   K.L.E. Kaiser and V.S. Palabrica, *Water Poll. Res. J. Canada*, 1991, **26**(3), 361-431.
4   G. Persoone, K. Wadhia and K. C. Thompson, 'Rapid toxkits microbiotests in water contamination emergencies' for *Water Contamination Emergencies : Can We Cope* eds., K.C. Thompson and J. Gray…, International Conference, Kenilworth, UK, 2004.

Websites
IRIS, http://www.epa.gov/iriswebp/iris/index.html
ExToxNet, http://ace.orst.edu/info/extoxnet/pips/qindex.html
Chemfinder, http://chemfinder.cambridgesoft.com

Test Kits
Eclox-M™ Water Test Kit, Capital Controls Limited, Didcot, Oxon, UK
DeltaTox™ PSI Test System, SDI Europe, Hook, Hampshire, UK
Chemets Cyanide Test, Galgo Limited, St Albans, Herts, UK
Berthold LB123AC α-β and LB1231 β-γ activity monitoring system, Berthold Technologies GmbH & Co.KG, Germany. UK Agent: Advanced Measurement Technology, Bracknell, Berkshire, UK

Shimadzu Multispec-1501 UV spectrophotometer, Shimadzu Deutschland GmbH (UK Branch), Milton Keynes, UK
PpbRAE Photo-Ionisation Detector monitor kit, Shaw City Limited, Faringdon, Oxfordshire, UK.
Uni-Lite® XCEL system, Biotrace Limited, Bridgend, UK
Thamnotox F™, MicroBioTests Inc, Nazareth, Belgium. UK Agent: ALcontrol Laboratories, Rotherham, S60 1BZ, UK.
Macherey Nagel Pesticide test kit, Duren, Germany

This work has been carried out mainly by ALcontrol as a R&D project financed by YWS. However, ideas have been contributed by both parties. Any views expressed are not necessarily the views of either company but those of the individual authors. No specific products are given the endorsement by either the authors or the companies.

MICROBIOLOGICAL ANALYSIS: HOW SOON CAN WE HAVE THE RESULTS?

David P. Sartory

Severn Trent Water, Quality and Environmental Services, Welshpool Road, Shrewsbury, SY3 8BJ, U.K.

1 INTRODUCTION

Ever since the pioneering work of John Snow and William Budd in the 19th Century demonstrating the link between contaminated water and outbreaks of cholera and typhoid[1,2] the need for testing water for contamination has been recognised. With the development of culture methods by Robert Koch, the first regular testing of water supplies was undertaken by Percy Frankland in 1885, who tested water from various parts of London and used a modification of Koch's procedure to assess the efficacy of water treatment by filtration.[3] With the greater understanding of enteric bacteria at the end of the 19th and in the early 20th Centuries, and the development of simple methods, the microbiological analysis of drinking water and allied samples became the cornerstone practice in assuring water quality and protecting public health. For much of the last 100 years, however, the methods that have been available for the detection or enumeration of indicator bacteria or pathogens have been relatively slow, as the standard methods employed for these organisms rely on their growth, the consequence of which is that it can take between 24 hours and 7 days before results are obtained. This is often too late for meaningful public health protection or for real-time monitoring of efficacy of water treatment and distribution.

Over the last twenty years there have been significant developments in the detection of bacteria through improved and specific substrates, immunological methods and gene detection. Some of these developments have resulted in reduced times between the collection of a sample and the availability of a result, especially for the classic bacterial indicators of water quality. For example, the use of fluorogenic or chromogenic enzyme specific substrates for *Escherichia coli* in media such as Colilert® QuantiTray™, an expanded most-probable-number (MPN) procedure that employs ortho-nitrophenol-β-D-galactopyranoside (ONPG) and 4-methylumbelliferyl-β-D-glucuronide (MUG) for the detection of β-galactosidase and β-glucuronidase respectively in a defined substrate medium[4], and membrane-Lactose Glucuronide Agar (m-LGA)[5] which is based upon membrane filtration and employs 5-bromo-4-chloro-3-indolyl-β-D-glucuronide (BCIG) as the marker for β-glucuronidase, have allowed confirmed counts of *E. coli* within 18 hours of onset of sample analysis. Some of these approaches, however, offer the promise of the results being available on the same day as collection of the sample, or even within a few hours, with several promising methods being published recently. Before such tests are employed, however, it is essential that they are appropriate for the task and situations where they are used.

2 THE REQUIREMENTS FOR RAPID METHODS

Microbiological testing of water supplies will typically be undertaken for either of two purposes, each of which have their own requirements for immediacy of availability of results and the organisms tested. It can be argued that for routine monitoring (whether for regulatory or operational purposes) a reliable count of an indicator bacterium is the key requirement, in that trends can be monitored or the degree of change assessed. For incident management, however, the need for a rapid indication of the significant presence of a pathogen is greater than an indication of the actual numbers present. In the first instance, therefore, the speed of analysis may be secondary to the acquisition of a numerical result, whilst in the second instance a quantitative result may be secondary to the need for a rapid result.

2.1 Rapid methods for routine monitoring

The key advantages of the methods currently used for operational monitoring are that they are relatively simple to perform and require unsophisticated laboratory equipment. They are also, unfortunately, labour intensive and require at least overnight incubation. However, a large number of samples (up to 500 or more in larger laboratories) can be analysed daily. A significant improvement for operational purposes would be achieved if sampling and analysis were completed on the same day, and early enough to allow operational or investigational actions to be taken that day. Any method that would provide results on the day of sample collection would need to meet some general requirements with respect to current capabilities and legislation before adoption. These are that the results should be available in 6 – 8 hours or less, that they are quantitative, and that there is close comparability with existing methods.[6]

Standards for indicator bacteria are legally defined and there are substantial databases of performance to these standards. Any change in methods that produces a significant increase in positive samples will cause considerable concern regarding the interpretation and significance of such results. There may a perception that water quality performance is deteriorating with respect to previous results due to the improved methodology but, arguably, without any increase in health risk.

There are some key attributes relating to performance, costs and support from the manufacturer that any same-day system would need to fulfil[7]:-

i)   Sensitivity and specificity – the system must be able to accurately detect the target organisms from high concentration of background and competing organisms. Sensitivity must be at least 1 organism per 100 ml. False-positives and false-negatives must be as close to zero as possible.

ii)   Speed – For operational or investigational actions to be undertaken within social hours (say before 8.00 pm) then results need to be available during the afternoon, preferably as early as possible. Since there may be restrictions on how early in the morning samples can be taken, then the time between collection of the sample and availability of results needs to be minimised, leaving a maximum of about eight hours for growth and detection.

iii)   Throughput – For the bigger microbiology laboratories a daily sample throughput of 200 - 500 samples is typical. Systems offering same-day results need to be able to handle this throughput without the need for banks of expensive detection equipment.

iv)   Non-destructive – There will always be a requirement to be able to subculture isolated bacteria for further study. The system must allow easy retrieval of such isolates.

v)   Analytical skills – Any system developed should not be so sophisticated as to require analysts with degrees or doctorates. Automated or semi-automated equipment may offer advantages in ease of use by analysts with basic or no microbiological training. Problem solving, however, may require more detailed knowledge.

vi)   Costs – The cost of initial purchase of the equipment and subsequent costs for reagents and materials must be commensurate with an acceptable cost per test in relation to current analytical costs. Expensive equipment may need to achieve very high throughputs per instrument compared to alternative cheaper equipment with lower throughputs per instrument.

vii)   Manufacturer's reliability and technical service – With a legal responsibility on water companies to achieve stipulated sample and analysis frequencies the manufacturers of equipment supplied to the water industry must have a very high level of reliability and technical back-up. The more complicated the equipment the more the laboratory is reliant upon the manufacturer's technical services.

2.2 Rapid methods for incident management

Generally the first criterion to the suspicion that a water supply may be implicated in the incidence of illness in a community is to establish whether the causative organism is or has been present in the supply. Thus one purpose of testing the water is to confirm or otherwise the water as the source, so that appropriate actions can be undertaken to prevent further illness and remove the organism from the supply. Further to this, testing aids health risk judgements and the application of any appropriate protective advice (e.g. advice to boil the water). Thus the faster a result can be made available, the better the risk assessment process, and the more timely and targeted any remedial action undertaken will be. In overall terms the criteria described above for routine monitoring[7] are applicable for incident management, but modified to take into account the different emphasis placed on the meaning of the result:-

i)   Sensitivity and specificity – As for routine monitoring the system must be able to accurately detect the target organisms from high concentration of background and competing organisms with a good degree of sensitivity. False-positives and false-negatives should not occur as far as possible, as key public health and operational decisions will made according to the results obtained.

ii)   Speed – For incident management an early meaningful result is important. This may mean that qualitative or semi-quantitative analyses may be more appropriate, and probably more applicable to analysis in the field.

iii)   Throughput – The numbers of samples taken in an incident may not be a significant increase for larger laboratories, but could be overwhelming for smaller facilities. To some extent this would be obviated if analyses could be undertaken in the field using simple test kits.

iv)   Non-destructive – This requirement may be sacrificed in yielding a rapid result. Parallel sampling and analysis using rapid and slower conventional non-destructive methods could be appropriate in confirming the presence of an organism and allowing further study if required (e.g. genotyping to confirm source and patient isolates are related).

v)   Analytical skills – For incident management, systems developed should be relatively unsophisticated, particularly if any analyses are to be undertaken in the field.

As before costs of the methods and manufacturer reliability must be commensurate with critical nature and circumstances under which the methods are used.

3 DEVELOPMENTS IN RAPID METHODS

There are basically two approaches to the rapid detection of microorganisms. The obvious approach is to attempt to detect the cells directly from water and allied samples. For some organisms like *Cryptosporidium*, *Giardia* and many viruses, this is the only choice currently available, but the methods for these organisms tend to require sophisticated concentration and detection equipment. For many bacteria, however, detection could be achieved more simply by using methods based upon conventional culture to generate more of the target organism prior to detection.

The future for conventional culture for rapid assessment of water quality has been recently reviewed[7]. The rapid detection or enumeration of bacteria by conventional culture will rely on a short period of replication (preferably no more than 4 hours), probably boosted by the addition of growth supplements, followed by detection of a signal which would be a demonstration of microbial growth, reflecting the viability of the bacteria present. This can be achieved by either membrane filtration techniques or conventional broth culture. Detection of microcolonies on a membrane filter could be carried out simply using laser scanning or high sensitivity cameras, pre-staining the colony with a suitable monoclonal antibody or other marker, providing the target is large enough for visualisation. Growth of target organisms into microcolonies also suggests that the samples would also have been positive by current conventional culture had incubation continued. The early detection of growth in broth culture could be by hand held detectors, provided the signal strength is generated relatively quickly and, if is to be quantitative, can be equated to extent of growth.

Detection and enumeration of microcolonies can be achieved via a variety of mechanisms. The use of substrates specific for target enzymes (e.g. β-galactosidase and β-glucuronidase for coliform bacteria and *E. coli*) which emit light when cleaved probably provides the quickest and easiest approach. A system based on this approach for the rapid enumeration of coliform bacteria and *E. coli* has been described.[8] The system is based on in-field processing of samples, incubation on a selective growth medium during transport to the laboratory, followed by the automated detection of microcolonies using epifluorescence microscopy. The microcolonies were visualised using the non-specific respiratory redox dye 5-cyano-2,3-ditolyl tetrazolium chloride (CTC), which required a reaction period once the sample had arrived at the laboratory, and the enzyme specific substrate 4-methylumbelliferyl-β-D-glucuronide (MUG) for β-glucuronidase (diagnostic for *E. coli*), which was incorporated into the selective growth medium. This approach resulted in *E. coli*, together with presumptive coliforms, being enumerated within 6 hours of collection of the sample. Subsequent development, including the derivation of novel fluorogenic substrates for both β-galactosidase and β-glucuronidase which yielded higher signal to background outputs, replacing the respiratory marker and MUG, allows the more specific enumeration of coliforms and *E. coli* within 5 to 5½ hours.[9] The issue of generating a high target fluorescence signal against a background fluorescence is important for systems using fluorogenic substrates for microcolony detection or fluorophores for detection of single cells, especially for analyses that will include a substantial amount of interfering autofluorescing material (e.g. in the detection of *Cryptosporidium* and *Giardia* from environmental samples). A recent development that helps address this problem is time-resolved fluorescence microscopy[10] whereby the target microcolony or cells are tagged with specific fluorophores that have a long lived fluorescence, which is still detectable after excitation illumination is switched off and autofluorescence has subsided. This approach has been successfully applied, using a conventional fluorescence microscope and an image-intensifying camera with a 60 μs delay between illumination and target detection, to the detection of *Giardia* in the presence of large amounts of autofluorescing material, resulting in a 30-fold increase in contrast of the labelled cysts compared to conventional immunostaining.[10]

An alternative to epifluorescence microscopy is laser scanning solid phase cytometry. Using the ChemScan® RDI system, Van Poucke and Nelis [11] have developed a two-step membrane filtration approach involving a limited period of 3 hours for growth in the presence of inducers for the production of specific enzymes, reaction with fluorogenic substrates and visualisation by laser scanning. This system is capable of enumerating single cells as well as any developed microcolonies. With this procedure *E. coli* could be enumerated from spiked water and naturally contaminated water samples within 3½ hours, with good correlations with conventional methods.[12] The authors concluded that the procedure would be useful for emergency monitoring of drinking water when rapid results are crucial.

Lux genes, typically from *Vibrio harveyi* and *Vib. fischeri*, regulate the production of bacterial luminescence and have been used to detect a range of bacteria.[13,14] This approach could be adapted to the rapid detection of microcolonies. Bacteriophages which have been genetically modified to bear these *lux* genes, and with the host cell lysis capability disabled, have been derived for a number of specific hosts including *E. coli*[15,16] and *Salmonella typhimurium*.[17] Procedures based on such *lux*-phage systems, or on specific substrates, would be non-destructive allowing further growth and subculture from any samples found to be positive for the target bacteria.

Other methods have been applied to the visualisation of microcolonies, but destroy the target cells during signal generation, in that they rely on cell lysis in order to allow reactions to take place. A 'rapid microorganism detection system' has been described[18] whereby coliforms could be detected after 6 to 7 hours incubation. Cellular ATP is extracted from the cells of the microcolonies, reacted with luciferin-luciferase reagents, and detected by an ultra-high-sensitivity camera.[19]

A gene probe approach to the visualisation of microcolonies was employed by Meier *et al.*[20] They derived fluorescently labelled rRNA targeted DNA probes for the detection by epifluorescence microscopy of species of *Enterococcus*. Some laboratory strains produced microcolonies after 4 to 5 hours incubation on a non-selective medium of sufficient size to be visualised by hybridisation within 8 hours. Enterococci from water samples could be detected within 20 hours. Combining chemiluminescence and gene probe technology, Stender *et al*[21] developed a chemiluminescent *in situ* hybridisation (CISH) method that allows the simultaneous detection, identification and enumeration of microcolonies of *E. coli* from drinking water. Samples are membrane filtered and incubated for 5 hours on a non-selective medium, after which *E. coli* microcolonies are visualised using a peroxidase-labelled peptide nucleic acid (PNA) probe targeting a species-specific sequence in *E. coli* 16S rRNA. Detection can be achieved using either Polaroid or X-ray film exposure or via a digital camera system, all which were successfully employed in the detection by CISH of *Pseudomonas aeruginosa* in bottled waters.[22] Analysis could be completed within a working day, and good correlations with conventional methods were achieved. An extension of this development is to use fluorescently-labelled PNA probes and laser or array scanners.[23] This latter approach would allow simultaneous detection and enumeration of a number of bacteria using probes with different fluorescent signatures.

An alternative approach to membrane filtration microcolony analysis would be to grow target organisms for a short period in a broth and then either detect growth directly from the broth culture or concentrate the resulting biomass for signal generation. Immunomagnetic separation techniques, using either immunomagnetic beads or ferrofluids, are now widely applied[24,25] and have been successfully used to recover *E. coli* O157 from water, either directly or following enrichment[26,27,28] and food,[29,26] *Yersinia enterocolitica* from water[30] and *Salmonella* from a range of foods,[31,32] as well as species of *Helicobacter* from centrifuged pellet concentrates from wastewater, groundwater and drinking water samples.[33] A range of immunomagnetic products for several pathogenic and indicator microorganisms is now commercially available. After immunomagnetic separation the recovered cells can be visualised using fluorogenic substrates and immunofluorescence,[26,34] chemiluminescence,[27] or polymerase chain reaction (PCR) and gene specific probes.[30,33] Although broth-based enrichment and immunomagnetic separation methods are more suited as part of the rapid determination of the presence of specific pathogens, this approach could be applied to recovery of indicator bacteria from water on a qualitative or semi-quantitative basis. To be quantitative an MPN format would be needed, which would probably make it too labour intensive for routine use, and would not offer any advantages over the microcolony formats discussed above. Savill *et al*[35] employed PCR for the specific detection of species of *Campylobacter* from recreational and drinking waters in New Zealand following MPN enrichment in Preston broth, which still took 72 hours for results to be available.

Lux-gene bearing bacteriophages could also be used to detect the growth of specific bacteria in broth culture. A *lux*-MPN method for the enumeration of *Sal. typhimurium* in environmental samples has been described.[36] By this method they were able to detect a single *Salmonella* cell from 100 ml of sewage sludge, soil or water after an overnight incubation. Although not same-day detection or enumeration the approach does reduce the analysis time to 24 hours as against 4 days by current conventional methods.

Chemiluminometric methods which generate a high light signal associated with growth or specific detection of a pathogen may offer a rapid detection system that is not reliant on sophisticated detectors. Van Poucke and Nelis[37] described a broth-based presence/absence (P/A) chemiluminometric method for detecting β-galactosidase which utilised a phenylgalactose-substituted 1,2-dioxetane derivative substrate. With this method they detected 1 coliform in 100 ml of sterile water after a 6 to 9 hour growth period. Problems were encountered, however, when the method was applied to natural waters, principally regarding its sensitivity.[38] Whilst there may be problems when trying to detect a heterogeneous group of bacteria, such as the coliform bacteria, this approach may have application for the rapid P/A testing for specific pathogens.

A procedure that incorporates some of the above approaches has been developed.[39] After a growth period of 5 to 6 hours, cells of either *E. coli* O157, *Sal. typhimurium* DT104 or *Sal. enteritidis* PT4 were recovered using immunomagnetic separation. The cells were then lysed using strain specific bacteriophages (after a further short growth period for *E. coli* O157), releasing endogenous adenyl kinase, which was detected using luciferin-luciferase bioluminescence. Using this procedure levels of 1 cfu of target bacterium in 25 ml of milk were detected within 8 hours. The use of selective immunomagnetic separation and the specificity of the bacteriophages used for lysis of the target cells means that this approach should be reliable in detecting the pathogen being sought.

Cultivation techniques can be very effective for bacteria, but have very limited application to the rapid detection of viruses or protozoan pathogens, as they generally need cell culture systems for replication,[40] although methods based upon such systems would be very advantageous in that they would also demonstrate the viability or potential infectivity of the pathogen detected.

The direct detection of single cells of a target pathogen or indicator bacterium is potentially very attractive in that it could remove the time needed for growth in the previous procedures. A multitude of fluorochromes and molecular probes which can be used detect the individual cell are now available, and have been widely applied to the assessment of total bacterial numbers in water. The DNA of cells can be stained using the binding fluorochrome 4-6-diamidino-2-phenylindole (DAPI) and then detected by flow cytometry or direct microscopy.[41,42] The assessment of physiological status of bacteria can be achieved using any of a wide range of fluorochromes currently available.[43] Redox dyes such as CTC and membrane potential dyes like rhodamine B can used to determine the viability of individual cells. CTC has been employed to determine total numbers of bacteria and regrowth in distribution systems.[41] Using a commercially developed viability dye, ChemChrome B, and a laser scanning detector[44] total bacterial numbers from drinking water could be obtained within 1 hour. These viability dyes, however, are not specific in determining the genus or species of any organisms being sought.

As with visualisation of microcolonies, single cells can be detected by a variety of methods. The use of labelled monoclonals against specific pathogens or faecal indicator bacteria together with a suitable viability dye has the potential of providing very rapid detection and enumeration methods. Filtration and staining would be simple and could be automated. Detection of single cells giving the appropriate signal, however, would require sophisticated detection equipment. By combining immunofluorescence using a fluorescein-labelled anti-O157 antibody with CTC viability staining Pyle *et al*[34] were able to detect *E. coli* O157:H7 in water by membrane filtration and epifluorescence microscopy within 3 to 4 hours, and similar procedures were applied to *Sal. typhimurium* and *Klebsiella pneumoniae*. However, the microscopy contributed significantly to the analytical time, making it labour intensive. The procedure was improved by recovering the cells using anti-O157 polyclonal antibody coated magnetisable beads, incubating the captured cells with CTC and staining them with a fluorescein-labelled antibody, followed by visualisation with the ChemScan® RDI laser scanning system.[26] With this procedure low levels of *E. coli* O157 were detected directly from food and water within 5 to 7 hours.

The nucleic acid-based techniques of polymerase chain reaction (PCR) and hybridisation probes for the specific detection of microorganisms from water is now well established,[45] particularly for the rapid identification of isolates. There is an abundance of methods now available for many species of bacteria, viruses and protozoa. Prescott and Fricker[46] described the use of PNA probes targeted against the V_1 region of *E. coli* 16S rRNA with visualisation using a CCD camera, which allowed the detection of *E. coli* in water in less than 3 hours. Lightfoot *et al*[47] using conventional PCR, were able to detect *Salmonella*, *Campylobacter* and *E. coli* O157 from spiked drinking water samples, concentrated by membrane filtration, within 6 hours, compared to the traditional cultural methods for these pathogens which take up to 4 days. However, the method was not as sensitive, with the limits of detection being around 10^3 organisms per unit volume sample analysed. The specificity of such molecular approaches does allow the simultaneous detection of more than one pathogen. Kong *et al*[48] have developed a multiplex PCR procedure, with separation and visualisation by gel electrophoresis and UV illumination of ethidium bromide tagged products, that is capable of detecting six bacterial pathogens (*Aeromonas hydrophila*, *Sal. typhimurium*, *Shigella flexneri*, *Yer. enterocolitica*, *Vib. cholerae* and *Vib. parahaemolyticus*) with a turnaround time of less than 12 hours, and detection limits of 1 to 10 cfu. The use of PCR and hybridisation probes for the rapid detection of pathogens, however, still suffers from an inability to differentiate between live and dead cells, questions of specificity and interference by other environmental material[45] although targeting nucleic acids that correlate with cellular activity, such as mRNA, may overcome some of these issues.

4   THE FUTURE FOR RAPID METHODS FOR MICROBIOLOGICAL WATER QUALITY ASSESSMENT

"In the dead of night, a storm rumbles across farmlands and cow pastures, torrents of rain drench the earth. Of the hundreds of cows that graze the saturated fields, several harbour E. coli O157:H7, a bacterium harmless to livestock but potentially deadly to

people. The bacterium, shed in the cow's manure, is washed into a stream that feeds a
public water district 20 miles away. Hours later, a red light flashes on an electronic
watershed map mounted on the control panel of the water district's monitoring station.
Microbial contaminants – detected by gene chips affixed to stationary stream posts and
inserted into wells – have entered the system. Other lights flash, indicating the identity of
the microbe. The lights alert the water district manager."

So begins a review on new tools for monitoring microbial water quality published in 2001
by the American Academy of Microbiology.[49] Whether such a system is realisable in the
near future is speculative, but acquiring results indicating the presence of a specific
pathogen within a few hours is already achievable. Over the last 20 years considerable
advancements in the detection of microorganisms from water and other environmental
materials have been made. An abundance of approaches has been described, many of
which are already applicable to rapid analysis of water samples, and others with potential
for application. Each approach has its advantages and limitations, which need to be
assessed when selecting one for application. The methods described in this paper
demonstrate that results for many procedures can be available within 4 to 6 hours, and
potentially earlier. This is already a considerable improvement on the current 18 hours
(for some indicator bacteria) to 72 hours or more (for some pathogens) using established
conventional methods. Future techniques may include gene-chip technology and biochips
with the potential for very rapid results.[45,49] In water quality assessment, however, the
target organism is often present in very low numbers and in the presence of a large
number of other bacteria, fungi, viruses and organic material, potentially complicating
any analytical procedure employed. For very rapid techniques to be applicable to water
analysis they will need to simultaneously address the concentration of the target
organism, its separation for co-concentrated material and its specific and reliable
detection. There is a wide range of techniques developed for each stage.[45,49] The trick is
to derive the combination of procedures that will deliver truly rapid quantitative or
qualitative methods, that allow as near real time monitoring and management of drinking
water supplies as possible. Rapid, same day testing, or indeed continuous monitoring,
would dramatically aid public health risk assessment and protection. Further focused
research in this area would bring significant benefits to drinking water quality
management and public health protection.

Several of the methods described here could be taken up with relatively few
problems, whilst others, such as PCR-based methods may only find limited application.
In the short term, methods based upon a short period of growth, particularly the
microcolony approach, offer the greatest likelihood of proving practical and acceptable to
microbiologists, health officials and regulators[7,45] at least for the bacterial indicators and
pathogens.

5 ACKNOWLEDGEMENTS

I am indebted to all my colleagues in water microbiology for many useful discussions
over the years, and to Martin Furness for useful comments on the manuscript. This paper

is published with the permission of Severn Trent Water, but the opinions expressed are those of the author and do not necessarily reflect those the company.

References

1 J. Snow J, *On the Mode of Communication of Cholera,* 1855. London: J. Churchill (Reprinted in *Snow on Cholera,* New York: Hafner Publishing Co. 1965).

2 W. Budd. (1873) *Typhoid Fever: Its Nature, Mode of Spreading and Prevention.* London: Longmans (Reprinted New York: American Public Health Association, 1931, available from http://www.deltaomega.org/classics.htm).

3 P. Frankland and G.C. Frankland, *Micro-organisms in Water: Their Significance, Identification and Removal.* London: Longmans, Green and Co, 1894.

4 E.G. Fricker, K.S. Illingworth and C.R. Fricker, *Water Research,* 1997 **31**, 2495 – 2499.

5 D.P. Sartory and L. Howard, *Letter in Applied Microbiology,* 1992, **15**, 273 – 276.

6 S.V. Sidorowicz and T.N. Whitmore, *Journal of the Institute of Water and Environment Management,* 1995, **9**, 92 – 98.

7 D.P. Sartory and J. Watkins, *Journal of Applied Microbiology Symposium Supplement,* 1999, **85**, 225S – 233S.

8 D.P. Sartory, A. Parton and C. Rackstraw, in *Rapid Detection Assays for Food and Water* (Eds. S.A. Clark, K.C. Thompson, C.W. Keevil and M.S. Smith), Cambridge: Royal Society of Chemistry, 2001, 31 – 37.

9 D.P. Sartory and A. Parton, Unpublished data

10 R. Connally, D. Veal and J. Piper, *FEMS Microbiology Ecology,* 2002, **41**, 239 – 245.

11 S.O. Van Poucke and H.J. Nelis, *Journal of Microbiological Methods,* 2000, **42**, 233 – 244.

12 S.O. Van Poucke and H.J. Nelis, *Journal of Applied Microbiology,* 2000, **89**, 390 – 396.

13 G.S.A.B. Stewart, *Letters in Applied Microbiology,* 1990, **10**, 1 – 8.

14 G.S.A.B. Stewart and P. Williams, *Journal of General Microbiology,* 1992, **138**, 1289 – 1300.

15 S. Ulitzur and J. Kuhn, In *Bioluminescence and Chemiluminescence: New Perspectives* (Eds. Schlomerich J., Andreesen R., Kapp A., Ernst M. and Woods W. G.), Chichester: John Wiley, 1987, 463 – 472.

16 C.P. Kodikara, H.H. Crew and G.S.A.B. Stewart, *FEMS Microbiology Letters,* 1991, **83**, 261 – 266.

17 G.S.A.B. Stewart, T. Smith and S. Denyer, *Food Science and Technology Today,* 1989, **3**, 19 – 22.

18 H. Tanaka, T. Shinji, K. Sawada, Y. Monji, S, Seto, M. Yajima and O. Yagi, *Water Research,* 1997, **31**, 1913 – 1918.

19 M. Masuko, S. Hosoi and T. Hayakawa, *FEMS Microbiology Letters, 1991,* **81**, 287 – 290.

20 H. Meier, C. Koob, W. Ludwig, R. Amann, E. Frahm, S. Hoffmann, U. Obst and K.H. Schleifer, *Water Science and Technology*, 1997, **35**, 437 – 444.

21 H. Stender, A.J. Broomer, K. Oliveira, H. Perry-O'Keefe, J.J. Hyldig-Nielsen, A. Sage and J. Coull, *Applied and Environmental Microbiology*, 2001, **67**, 142 – 147.

22 H. Stender, A. Broomer, K. Oliveira, H. Perry-O'Keefe, J.J. Hyldig-Nielsen, A. Sage, B. Young and J. Coull J, *Journal of Microbiological Methods*, 2000, **42**, 245 – 253.

23 H. Stender, K. Oliveira, S. Rigby, F. Bargoot and J. Coull, *Journal of Microbiological Methods*, 2001, **45**, 31 – 39.

24 I. Safarik, M. Safariková and S.J. Forsythe, *Journal of Applied Bacteriology*, 1995, **78**, 575 – 585.

25 T.N. Whitmore S. and Sidorowicz, *Microbiology Europe*, 1995, **3**, 16 – 22.

26 B.H. Pyle, S.C. Broadway and G.A. McFeters, *Applied and Environmental Microbiology*, 1999, **61**, 1966 – 1972.

27 D.R. Shelton and J.S. Karns, *Applied and Environmental Microbiology*, 2001, **67**, 2908 – 2915.

28 Standing Committee of Analysts, The Microbiology of Drinking Water 2002 – Part 4 – Methods for the Isolation and Enumeration of Coliform Bacteria and *Escherichia coli* (including *E. coli* O157:H7). *Methods for the Examination of Waters and Associated Materials*, London: Environment Agency, 2002.

29 D.J. Wright, P.A.Chapman and C.A. Siddons, *Epidemiology and Infection*, 1994, **113**, 31 – 39.

30 G. Kapperud, T. Vardund, E. Skjerve, E. Hornes and T.E. Michaelsen, *Applied and Environmental Microbiology*, 1993, **59**, 2938 – 2944.

31 L.P. Mansfield and S.J. Forsythe, *Letters in Applied Microbiolog,y* 1993, **16**, 122 – 125.

32 C. Poppe, L.A. Elliott and C.L. Duncan, *Journal of Microbiological Methods*, 1996, **25**, 237 – 244.

33 K. Hultén, H. Enroth, T. Nyström and L. Engstrand, *Journal of Applied Microbiology* 1998, **85**, 282 – 286.

34 B.H. Pyle, S.C. Broadway and G.A. McFeters, *Applied and Environmental Microbiology*, 1995, **61**, 2614 – 2619.

35 M.G. Savill, J.A. Hudson, A. Ball, J.D. Klena, P. Scoles, R.J. Whyte, R. McCormick and D. Jankovic, *Journal of Applied Microbiology*, 2001, **91**, 38 – 46.

36 P.E. Turpin, K.A. Maycroft, J. Bedford, C.L. Rowlands and E.M.H. Wellington, *Letters in Applied Microbiology*, 1993, **16**, 24 – 27.

37 S.O. Van Poucke and H.J. Nelis, *Applied and Environmental Microbiology*, 1995, **61**, 4505 – 4509.

38 S.O. Van Poucke and H.J. Nelis, *Applied and Environmental Microbiology*, 1997, **63**, 771 – 774.

39 K.J. Brown, A.J. Corbitt, J.A. Archard, M.J. Murphy, R.L. Leslie and D.J. Squirrell Poster presentation at the conference on Pathogens in the Environment and Changing Ecosystems, Society for Applied Microbiology, University of Nottingham, 8 – 11 July 2002.

40 IWA (International Water Association) (2000) Rapid microbiological monitoring methods: the status quo. IWA Standards and Monitoring Specialist Group task force report. Available from www.iwahq.org.uk/pdf/bp0002.pdf.

41 P. Cervantes, V. Mennecart, C. Robert, M.R. de Roubin and J.C. Joret, In *The Microbiological Quality of Water* (Ed. Sutcliffe D. W.) Ambleside: Freshwater Biological Association, 1997, 54 – 62.

42 J. Watkins and J. Xiangrong, In *The Microbiological Quality of Water* (Ed. Sutcliffe D. W. Ambleside: Freshwater Biological Association, 1997, 19 – 27.

43 G.A. McFeters, F.P. Yu, B.H. Pyle and P.S. Stewart, *Journal of Microbiological Methods,* 1995, **21**, 1 – 13.

44 D.T. Reynolds and C.R. Fricker, *Journal of Applied Microbiology,* 1999, **86**, 785 – 795.

45 T. Whitmore, *Water and Waste Treatment*, 2002, **45(6)**, 15 – 16.

46 A.M. Prescott and C.R. Fricker, *Molecular and Cellular Probes,* 1999, **13**, 261 – 268.

47 N. Lightfoot, M. Pearce, B. Place and C. Salgado, in *Rapid Detection Assays for Food and Water* (Eds. Clark S. A., Thompson K. C., Keevil C. W. and Smith. M. S.), Cambridge: 2001, Royal Society of Chemistry, 59 – 65.

48 R.Y.C. Kong, S.K.Y. Lee, T.W.F. Law, S.H.W. Law and R.S.S. Wu, *Water Research,* 2002 **36**, 2802 – 2812.

49 J.B. Rose and D.J. Grimes, *Reevaluation of Microbial Water Quality: Powerful New Tools for Detection and Risk Assessment.* A report of the American Academy of Microbiology. Washington: American Academy of Microbiology, 2001.

RAPID TOXKIT MICROBIOTESTS FOR WATER CONTAMINATION EMERGENCIES

G. Persoone[1], K. Wadhia[2] and K. Clive Thompson[2]

[1]Ghent University, Laboratory of Environmental Toxicology and Aquatic Ecology
J.Plateaustraat 22, B-9000 Ghent, Belgium
[2]ALcontrol Laboratories, Templeborough House, Mill Close, Rotherham S60 1BZ, England

1 INTRODUCTION

Water contamination is a very general term that refers to the accidental as well as the deliberate introduction of undesired and/or harmful agents in surface waters or groundwaters, and by extension into water supplies. Water contamination can originate from a variety of different sources, encompassing microbiological, virological, biological as well as physical/chemical agents.

Physical/chemical contaminants can impart an unpleasant appearance, taste or smell to the water which can trigger an immediate the alarm. Microbiological, virological, biological and most toxin contaminants in contrast, are in most cases only detected after an effect has occurred to environmental biota and/or man.

Irrespective of the type of incident, it is imperative that toxicity is detected and preferably also quantified "as rapidly as possible" to determine its nature and the extent of the hazard.

This paper focuses on the toxicological methods currently available for detection and quantification of the hazards concerning chemical contamination of water.

2 CHEMICAL ANALYSIS VERSUS TOXICITY TESTING

The provocative use of the terms chemistry and toxicology expressed in juxtaposition in the heading of this section is deliberate since there is a need to underline the continuing confusion of both these types of analysis.

Chemical analyses provide information on the nature and quantity of the chemical(s) present in a sample, but do not give evaluation of the effects that may result to the biota. Toxicity tests reveal the measure of impact of the contaminants on living organisms giving no insight into the nature or quantity of the relevant chemical(s) present.

The two approaches are in fact necessary and complementary to determine "causes" as well as "effects". However, both chemical analyses and toxicity tests have intrinsic shortcomings and weaknesses.

With regard to the chemical approach for water emergencies, the very first question is indeed "what to search for"? The array of chemicals which may have contaminated a particular water is indeed quite substantial" and hence would require an array of (sophisticated and expensive) apparatus and techniques to cope with all the categories of inorganic and organic compounds.

The determination of specific chemicals is complex and time consuming, and the results may not be available for hours.

Although a variety of chemical test kits are presently available with which toxicants can be rapidly identified, even in the field, their disadvantage is that they are "chemical specific" and hence only deal with one compound or group of compounds[1]. Sometimes these give poor results at low ambient temperatures. The same drawback also applies to immunoassays such as the well-known ELISA tests, which are commercially available but which are also highly "chemical specific".

Toxicity tests are an appropriate tool to employ to obtain a (biological) signal for harmful chemical(s) that may be present in contaminated water, but these also have their limitations. Firstly, the test species used must be sufficiently sensitive to give an effect signal. Since toxicity is "species as well as chemical specific", this means that in fact the level of the contamination must be sufficiently high to obtain an effect! Secondly, the exposure time must be long enough for the organisms to react to the toxicant(s); an important consideration that will be discussed in more detail below.

Finally, toxicity tests are dependent on the "availability" of living biological material, which means that live stocks of the selected species must be cultured year-round to ensure that in case of an incident, emergency testing can be performed 365 days a year. As emphasized in various papers[2-5], the facilities and especially the costs of the continuous culturing/maintenance of test organisms (to date) is still the major bottleneck in ecotoxicology, which restricts toxicity testing to a low number of highly specialized laboratories.

3 MICROBIOTESTS

The awareness of the biological, technical and not the least the financial burden of "conventional" bioassays, and in addition the need for low-cost alternatives triggered the development of alternative assays, and the appearance of "microbiotests".

As described by Blaise[6], microbiotests are low-cost and user-friendly small-scale assays which are either based on small organisms that are easy to culture or preferably culture/maintenance free, and which have the high sample throughput potential needed for routine testing. The same author, however, underlined that the former characteristics of microbiotests should not lead to any loss of precision nor lower sensitivity in comparison to conventional assays.

After the appearance of the very first "fully culture/maintenance independent" microbiotest with lyophilised bacteria[7], it took about a decade before standardized small-scale assays with aquatic invertebrates became available[8].

Eventually a whole range of microbiotests with various aquatic biota (micro-algae, ciliated protozoans, rotifers and crustaceans) was developed in the Laboratory for Biological Research in Aquatic Pollution (presently renamed Laboratory of Environmental Toxicology and Aquatic Ecology) at the Ghent University in Belgium. All these bioassays, which received the generic name "Toxkits", are based on biological stages of the test biota which are either, inactive, dormant or immobilized, which can be stored for several months to several years, and which can be "revitalized" at the time of performance of the toxicity tests [9-12].

These simple, user-friendly and low cost Toxkit microbiotests have been commercialised and are presently in use worldwide for routine toxicity detection and biomonitoring, and their various applications continue to be reported in increasing number of scientific publications; the titles and abstracts of these can be found on the website www.microbiotests.be

3.1 Application of Microbiotests for Water Contamination Emergencies

Although at first sight there is no direct relationship between the effects found with non-vertebrate species (prokaryotic or eukaryotic) and those which may occur in man, ecotoxicological tests are nevertheless in use to date to indicate the presence of harmful compounds in contaminated waters, as early warnings for impacts on humans[13].

Toxkit microbiotests have already proven their value as screening sentries for biotoxins produced by blue-green algae in water reservoirs[14-20]. The Thamnotoxkit F e.g. appears to be as sensitive as the mouse assay in detecting cyanotoxins and as a result this Toxkit assay is now used in several countries as an attractive low cost tool for routine applications at places where heavy blooms of cyanobacteria regularly occur. Work on response to other toxins is in progress.

3.2 Microbiotests with Invertebrates for Water Contamination Emergencies

In cases of real emergencies, i.e. those where answers are needed in a few hours, Toxkit microbiotests based on invertebrate test species are unfortunately not of any help, at least when the prescribed test procedures are applied. These small-scale assays are based on "mortality" as the effect criterion, with exposure times similar to those used in conventional toxicity tests, i.e. at least 24 hours...

As shown in Figure 1, there is, however, a minimum time necessary to obtain an effect that is independent of the concentration of the toxicant to get an effect, irrespective of the exposure time. Concerning mortality, the effect criterion is in many cases of water contamination incidents not sensitive enough to give a rapid signal, i.e. within a few hours.

Figure 1   *Time/Concentration relationship for toxic effects of chemicals on living organisms*

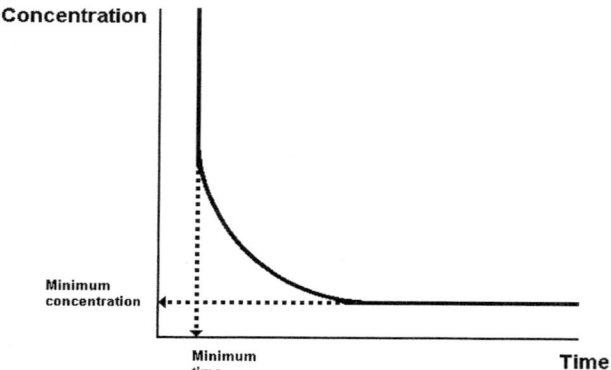

4 MICROBIOTESTS WITH SUB-LETHAL EFFECT CRITERIA

All living biota (including man) are, however, already affected (potentially suffering) when exposed to harmful agents at levels substantially below "lethal" concentrations. Chronic toxicity tests have hence been developed whereby the test biota are exposed to the toxicants for long periods (several days to several weeks) to detect the impact of low concentrations on growth and reproduction. These assays, which are much more sensitive than "acute (mortality) tests" are unfortunately not appropriate for emergency situations because of the long exposure times.

The presence of harmful substances can often be detected rapidly by measurement of their impact on various biological activities of biota: - physiological, metabolical or even behavioural.

The best known and most applied tests in this regard are undoubtedly those which measure enzymatic inhibition in bacteria after exposure times as short as 15 minutes. Several bioluminescence tests with lyophilised marine bacteria (mostly *Vibrio fisheri*) are to date available commercially such as Microtox™ – Deltatox™ – Lumistox™ – ToxAlert™ - CheckLight Toxscreen™, BioTox™, and in use worldwide for rapid detection of toxicity in water samples. The bacterial luminescence inhibition tests are undoubtedly the best-documented microbiotests and the number of scientific publications reporting their applications exceeds several hundred.

Irrespective of the merits of these very rapid bacterial assays, their sensitivity is unfortunately very chemical specific and as shown in Table 1[13] the EC_{50}s range from very low to very high concentration depending on the type of compounds.

Table 1   *Comparison of the sensitivity of Microtox™ and Thamnotoxkit F™ for various chemicals (all results expressed in L(E)C50s)*

Chemical	Microtox™ (15 min)	Thamnotoxkit F™ (24 h)
Arsenic (III)	0.3	1.3
Cyanide (KCN)	3.5	0.3
Rat poison a	507	0.4
Dichlorvos	29	19
Thallium	450	0.2
Paraquat	20	0.6

Reference: O'Neill, G. et al, Chapter 13 of this volume.

Enzymatic assays based on measurement of the inhibition of galactosidases or esterases are now also available with aquatic invertebrates. The 1h Fluotox microbiotest[21-23] e.g. is based on the decrease of galactosidase activity in the test species, with visual observation of decreased fluorescence. The organisms are first exposed for one hour to the toxicant, after which time a complex substrate composed of a fluorescent part and a galactoside is added. If the galactosidases are inhibited, the ingested substrate is not cleaved and the (originally bound) fluorescing part is not set free, so that the biota do not become fluorescent. The data generated with several types of test species and several types of compounds revealed that the 1 h enzymatic inhibition data correlated very well with 24h mortality figures.

Rapid toxicity tests have also been developed recently based on behavioural endpoints such as swimming and feeding. The swimming behaviour of the rotifer *Brachionus calyciflorus* exposed to toxicants for periods of minutes up to a few hours was found to be affected at concentrations lower than the 24 h LC_{50}s[24]. This microbiotest was subsequently automated by computer-aided video image capture[25] showing the potential of behavioural endpoints in toxicity assessment.

Inhibition of food ingestion under stress is a well-known ecological phenomenon, the potential of which has been explored for rapid toxicity testing. Screening methods based on decreased food uptake (micro-algae) by rotifers and crustaceans exposed for a few hours to toxicants have been described by Ferrando et al[26-27].

Interesting variants of the latter methods have been developed concurrently by various scientists, with replacement of live foods by inert foods. The 1h Ceriofast assay with the crustacean *Ceriodaphnia dubia*[28] measures the reduction of uptake of yeast cells stained with a fluorescent dye. Juchelka and Snell[29-30] advancing developments used fluorescently labelled microspheres to measure the reduction of particle uptake in ciliates, rotifers and crustaceans.

5 SUB-LETHAL TOXKIT MICROBIOTESTS BASED ON PARTICLE UPTAKE

The discovery that many aquatic invertebrates are "non-selective filter feeders" and ingest particles irrespective of their vitality and source providing that they conform to a specific size range, triggered research in the company Microbiotests Inc. to adapt the previously marketed technologies and develop "rapid Toxkit microbiotests" based on particle uptake.

Experiments were performed with *Brachionus calyciflorus* (Rotoxkit F™) and *Thamnocephalus platyurus* (Thamnotoxkit F™) to determine the uptake of coloured (red) latex spheres of 5 μm diameter and of activated carbon particles, after the organisms had been exposed for various periods (minutes to hours) to different concentrations of toxicants.

The investigations revealed that within minutes of administering the particles, the guts and the digestive tract of the test species that had not been exposed to toxicants, were full of particles, which can easily be seen under a dissection microscope at low magnification (see Figure 2). When exposed to toxicants, there was less particle ingestion and even complete lack of ingestion, depending on the time of exposure and the toxicant concentration (see Figure 3).

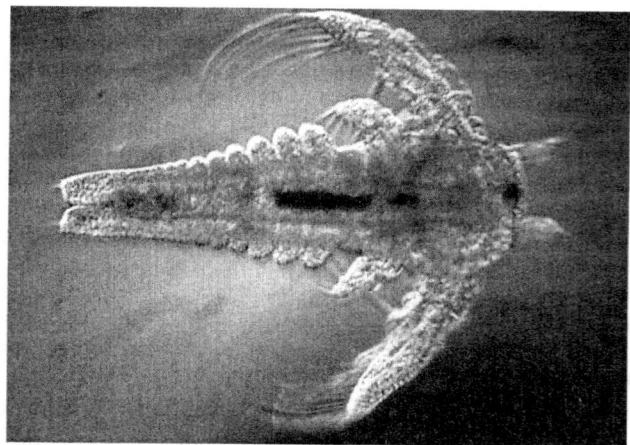

Figure 2   *Microphotograph of Thamnocephalus platyurus larva (400 μm) left for 1h in a non- toxic medium, followed by addition of activated carbon. The photo clearly shows the uptake of the black particles in the digestive tract.*

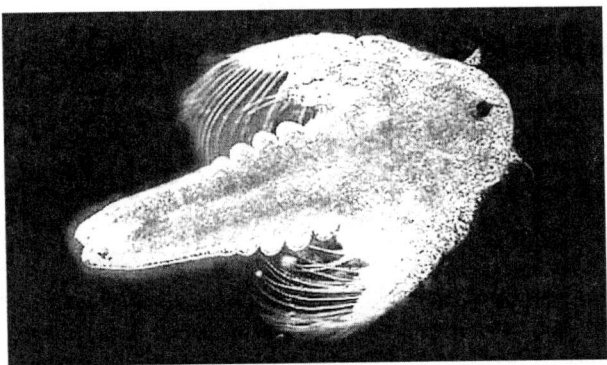

Figure 3   *Microphotograph of Thamnocephalus platyurus larva (400 μm) exposed for 1h to a toxic solution, followed by addition of activated carbon. The photo shows that the digestive tract is empty and that the organisms show lack of ingestion of the particles.*

The experiments performed with the two types of particles subsequently revealed that although both the latex particles and the activated carbon were taken up by the organisms and visible in the gut and the digestive tract, the (black) activated carbon was nevertheless more suited for visual observation in the crustacean test species, whereas the red latex particles were more suited for the rotifer assays.

On the basis of the preliminary findings, a very simple procedure was worked out for the rapid particle ingestion test: 1h exposure of the test species to the toxicant followed by addition of the particles and then microscopic observation 15 minutes later.

The endpoint which is taken into consideration for the interpretation of the results of tests performed with a dilution series of a toxicant, is the lowest concentration at which there is a (visually) significant lower uptake of particles in the test organisms than in those of the controls. In ecotoxicology this concentration is referred to as the LOEC (Lowest Observed Effect Concentration), whereas the next lowest concentration is the NOEC (No Observed Effect Concentration). In cases of water contamination emergencies, it suffices to compare the amount of particles taken up by the controls versus those in the suspected sample, and decide whether or not there is "significant" difference between the two.

The crucial question with regard to the potential of the new particle ingestion microbiotests is, however, whether or not these rapid tests will be sensitive enough to signal the presence of hazardous compounds through lower uptake of particles in the suspected water sample compared with that in control water.

Although no conclusive answer can be given at this time, a preliminary series of experiments performed with the rotifer and crustacean test species named above, with the reference chemicals potassium dichromate ($K_2Cr_2O_7$) and sodium pentachlorophenolate (NaPCP), revealed that the 1h particle uptake assay was as sensitive as the 24h lethality test with the same test biota. The comparison made by O'Neill et al (Chapter ?) between the EC_{50}s of the rapid bacterial luminescence test (Microtox™) and the 24h LC_{50} of the Thamnotoxkit furthermore indicate that for the majority of the chemicals reported, the invertebrate microbiotest was substantially more sensitive than the bacterial assay. From both these facts, and in expectation of additional data, it seems that the new rapid particle ingestion Toxkit microbiotests have a significant potential for application in emergency situations.

6 A FINAL HURDLE…

Earlier in this paper, it was indicated that the major handicap in routine ecotoxicological testing and biomonitoring is the continuous availability of stocks of live test species, a bottleneck that has now been resolved owing to the availability of the "culture/maintenance independent" microbiotests.

However, unlike the bacterial tests, for which the (lyophilised) material can be "revitalised" in a few minutes, the microbiotests based on invertebrate tests species do require one to several days for completion of the embryonic development and the hatching of the test biota from the dormant stages… For most applications of toxicity tests there is usually no need for "immediate" testing and this delay in obtaining the biological material for the assays is not crucial.

In cases of water contamination emergencies, the live test organisms must be available "on the spot", seven days a week, to cope with incidents…

This last hurdle is presently being addressed by the company Microbiotests Inc. Progress has been made with the design of an "automatic hatching incubator"; this is in the final stage of development. The apparatus will provide "freshly hatched" test species on a daily basis, with which rapid particle uptake microbiotests can be performed at any time. This overcomes the need to manually implement the hatching process on an on-going basis.

7 CONCLUSIONS

Chemical analyses and toxicity tests are complementary tools that have their respective merits to cope with emergency situations of water contamination.

Rapid "culture/maintenance free" toxicity tests with bacteria are now available, but their sensitivity is highly "chemical dependent". Consequently the general ecotoxicological principle that a (small) battery of tests with different species is necessary to cope with different types of contaminants is also valid in the case of water contamination incidents.

Rapid Toxkit microbiotests with selected aquatic invertebrates, based on enzymatic or particle ingestion effect criteria have recently been developed, which can fulfil this role. Their application in laboratories dealing with emergency situations is, however, dependent of the development of specific equipment to daily provide the live biological material from the dormant eggs.

References

1   M. D. Buck, 'Rapid chemical assays based on test kits' in *Rapid Detection Assays for Food and Water*, eds., S.A. Clark, K.C. Thompson, C.W. Keevil and M.S. Smith, Royal Society of Chemistry, Special Publication N° **272**, 2001, 73-79.

2   C. R. Janssen and G. Persoone, 'Routine aquatic toxicity testing: some problems and new approaches' in *Biological Indicators of Environmental Monitoring*, eds., S. Bonotto, , R. Nobili and R. Revoltella, Serano Symposia Review 27, Rome, 1991, pp. 195-207.

3   G. Persoone, *Zeitschr. für Angew. Zool.*, 1992, **78(2)**, 235-241.

4   C. Janssen, 'Alternative assays for routine toxicity assessments : a review' in *"Ecotoxicology"*, eds., G. Schüürmann and B. Market, John Wiley and Sons, 1998, Chapter 26, 813-839.

5   G. Persoone, 'Microbiotests for rapid and cost-effective hazard assessment of industrial products, effluents, wastes, waste leachates and groundwaters' in *Rapid Detection Assays for Food and Water*, eds., S.A. Clark, K.C. Thompson, C.W. Keevil and M.S. Smith, Royal Society of Chemistry, Special Publication N° **272**, 2001, 73-79.

6   C. Blaise, *Environ. Toxicol. Water Qual.*, 1991, **6**, 145-155.
7   A.A. Bulich, 'Use of luminescent bacteria for determining toxicity in aquatic environments' in *Aquatic Toxicology*, eds., L.L. Markings and R.A. Kimerle, ASTM, 1979, **667**, 98-106.
8   T.W. Snell. and G. Persoone, *Aquatic Toxicology*, 1989, 14, p. 81-92.
9   M.D. Centeno, L. Brendonck and G. Persoone, 'Cyst-based Toxicity Tests : III. - Development and standardization of an acute toxicity test with the freshwater anostracan crustacean *Streptocephalus proboscideus*' in *Progress in Standardization of Aquatic Toxicity Tests* eds.,. A.M.V.M. Soares and P. Calow, Lewis Publishers, 1992, 37-55.
10   M. Van Steertegem and G. Persoone, 'Cyst-based Toxicity Tests : V. - Development and critical evaluation of standardized toxicity tests with the brine shrimp Artemia (Anostraca, Crustacea)' in *Progress in Standardization of Aquatic Toxicity Tests*, eds., A.M.V.M. Soares and P. Calow, Lewis Publishers, 1992, 81-97.
11   G. Persoone, 'Development and first validation of a "stock culture free" algal microbiotest : the Algaltoxkit' in *Microscale Aquatic Toxicology, Advances, Techniques and Practice*, eds., P.G. Wells, K .Lee and C. Blaise, CRC Lewis Publishers, Chapter 20, 1998, 311-320.
12   G. Persoone, 'Development and validation of Toxkit microbiotests with invertebrates, in particular crustaceans' in *Microscale Aquatic Toxicology, Advances, Techniques and Practice*, eds., P.G. Wells, K .Lee and C. Blaise, CRC Lewis Publishers, Chapter 30, 1998, 437-449.
13   G. O'Neill, C. Ridsdale, K.C. Thompson, and K. Wadhia, 'Field and laboratory analysis for detection of unknown deliberately released contaminants' in *Water Contamination Emergencies : Can We Cope* eds., K.C. Thompson ..., International Conference, Kenilworth, UK, 2003, ..
14   A. Törökne, *Environ.Toxicol.*, 1999, **14**, 5, 466-472.
15   A. Törökne, 'The potential of the Thamnotoxkit microbiotest for routine detection of cyanobacterial toxins' in *New Microbiotests for Routine Toxicity Screening and Biomonitoring* eds., G. Persoone, C. Janssen and W. De Coen, Kluwer Academic/Plenum Publishers, 2000, 533-540.
16   B. Marsalek and L. Blaha, 'Microbiotests for cyanobacterial toxins screening' in *New Microbiotests for Routine Toxicity Screening and Biomonitoring* eds., G. Persoone, C. Janssen and W. De Coen, Kluwer Academic/Plenum Publishers, 2000, 519-526.
17   M. Tarczynska, G. Nalecz-Jawecki, M. Brzychey and J. Sawicki, 'The toxicity of cyanobacterial blooms as determined by microbiotests and mouse assays' in *New Microbiotests for Routine Toxicity Screening and Biomonitoring* eds., G. Persoone, C. Janssen and W. De Coen, Kluwer Academic/Plenum Publishers, 2000, 527-532.
18   A. Törökne, E. Laszlo, I. Chorus, J. Fastner, R. Heinze, J. Padisak and F. Barbosa, *Water Science & Technol.*, 2000, **42**, 1-2, 381-385.
19   A. Törökne, E. Laszlo, I. Chorus, K. Sivonen and F. Barbosa, *Environ.Toxicol.*, 2000, **15**, 5, 549-553.
20   A. Törökne, O. Reskone and N. Baskay, *Central European Journal of Public Health*, 2000, **8**, 97-99.
21   C.R. Janssen and G. Persoone, *Environmental Toxicology and Chemistry*, 1993, **12**, 711-717.
22   C.R. Janssen, E. Espiritu and G. Persoone, 'Evaluation of the new "enzymatic inhibition" criterion for rapid toxicity testing with Daphnia magna'. pp. 71-81. in *Progress in Standardization of Aquatic Toxicity Tests*, eds., A.M.V.M. Soares and P. Calow, Lewis Publishers, 1992, 71-81.

23 E. Espiritu, C.R. Janssen and G Persoone, *Environmental Toxicology and Water Quality*, 1995, **10**, 25-34.

24 C. R. Janssen, M.D. Ferrando and G. Persoone. *Ecotox. Env. Safety*, 1994, **28**, 247-255.

25 C. Charoy, C.R. Janssen, G. Persoone and P. Clement, *Aquatic Toxicology*, 1995, **32**, 271-282.

26 M.D. Ferrando, C.R. Janssen, E. Andreu and G. Persoone, *Ecotoxicology and Environmental Safety*, 1993, **26**, 1-9.

27 M.D. Ferrando, C.R. Janssen, E. Andreu and G. Persoone, *Sci. Total Environ.*, 1993, 1059-1069.

28 G. Bitton, K. Rhodes and B. Koopman, *Env. Toxicol. Chem.*, 1996, **15**, 2, 123-125.

29 C.M. Juchelka, and T.W. Snell, *Archiv. Environ. Contam.Toxicol.*, 1994, **26**, 549-554

30 C. M. Juchelka. and T.W. Snell, *Archiv. Environ. Contam .Toxicol.*, 1995, **28**, 508- 512

OVERVIEW OF HANDLING EMERGENCIES: SOME CAUTIONARY TALES

R.A. Deininger

School of Public Health, The University of Michigan, Ann Arbor, MI 48109

1   INTRODUCTION

Water supply systems are vulnerable to natural and man-made disasters. They are vulnerable to destruction by tornadoes and earthquakes, and an intentional destruction of key elements of a supply system is possible through the use of explosives. The electrical supply network necessary for the operation of pumps and the treatment processes may fail and lead to a loss of the delivery of water to the consumers. Pumps and pipes break and lead to a partial shutdown of the system. Over the years utilities have learned to adjust to such emergencies and have procedures and materials in place to recover from such accidents. A determined attack by a group of terrorist may cause havoc and the disruption of service. How to counter such threats, and what emergency procedures should be taken, is underway for every water treatment system in the US that serves more than 3,300 customers through mandatory vulnerability assessments. A contamination of the system, unexpected or deliberate, is of very serious concern since our detection systems are not very good or non-existent, and the consequences of death or illness may cause a panic in the population. No successful large scale attacks on US water supplies have happened, so it is instructive to look at accidental outbreaks of "natural" contamination of systems and study what happened, what was done, and what could have been done to prevent such an event.

I have somewhat arbitrarily chosen 5 outbreaks of waterborne disease and will discuss them in detail in the following:

> Milwaukee, Wisconsin, *Cryptosporidium* outbreak
> North Battlford, Saskatchewan *Cryptosporidium* outbreak
> Walkerton, Ontario, *E. coli O157* outbreak
> Gideon, Missouri., *Salmonella* outbreak
> Detroit, Michigan, main break, no outbreak

There are lessons to be learned from these outbreaks and they will be discussed.

2   CASE EXAMPLES

The first two examples chosen deal with the outbreak of cryptosporidiosis caused by the parasite *Cryptosporidium parvum*. The next two deal with the outbreak from bacteria. And the last example caused no illness, just massive disruption of service. There are lessons to be learned from each example.

2.1   The Milwaukee, Wisconsin, cryptosporidiosis outbreak

The largest outbreak of a waterborne disease in the U.S. occurred in the spring of 1993 in Milwaukee, Wisconsin. The water supply system of Milwaukee serves about 800,000 people from two treatment plants that draw their water from Lake Michigan through two cribs located about 2 km offshore at a depth of over 10 meters.. The papers reported that over 400,000 people became ill, and that 100 people died. Subsequent analysis showed that these numbers are a bit overstated, but even if one cuts these numbers into half, the outbreak is still the largest.

In March there were heavy rains, and thawing of the ice accumulated during the winter months, so that there was heavy storm water runoff and runoff from the agricultural lands. The Milwaukee water supply system has two water plants, one in the north called the Linwood plant (LWTP) that serves the northern part of the city, and one in the South called the Howard Avenue Water Plant (HWTP) that serves generally the southern part of the city. There is a part of the city where the source of the supply varies, sometimes from the north, sometimes from the south. During March the water consumption is low enough so that one plant can provide enough water. The Milwaukee River that carries the discharge of the storm sewers discharges into Lake Michigan in between the two plants. The near-shore currents in the lake are predominantly southern so that the plume of waste water passed over the southern intake of the HWTP.

On April 6 the diagnosis of cryptosporidiosis was made in the first patient, and the number of patients climbed rapidly. Residents of nursing homes in the southern part had an over 10 times higher incidence of gastrointestinal illness than those in a northern nursing home (Fox[4]). The Howard street water plant was shut down, and a general boil water order was issued.

A review of the operating data of the HWTP showed that maintenance work was performed on the filters, and that the turbidity of the effluent from the filters increased from around 0.5 ntu to over 2.0 ntu during late March and early April. Much closer attention should have been paid to these parameters, and the dosages of the coagulants could have been adjusted..

In the end the city decided to relocate the intake further out into Lake Michigan, to improve the filters, and to install an ozonation process as primary disinfectant. It is this multi -barrier approach that should keep the water supply of Milwaukee safe.

2.2   The North Battleford, Saskatchewan, cryptosporidiosis outbreak

In the spring of 2001 the City of North Battleford, Saskatchewan, experienced an outbreak of cryptosporidiosis. Out of the total population of about 15,000 roughly 7,000, or about half of the population, became ill. There are no reported deaths from the outbreak. Numerous visitors to the town also became ill.

The city is served by two water treatment plants both drawing water from the Saskatchewan River. One is a ground water plant upstream of the city with wells located near the river. This plant might be better termed a river bank filtration system, since the wells are directly influenced by the river water. The other plant is downstream from the city and takes water directly from the river through an intake. This second plant was originally built in 1950 to serve a mental hospital downstream from the city, and from a description of its many processes it was a museum of water treatment processes.

The city wastewater is treated in an activated sludge treatment plant with limited capacity and discharges into a channel that enters the river about 2 km upstream of the surface water treatment plant. Storm water is discharged directly to the channel, and so are waste waters from a hog processing plant that has some lagoon treatment.

When it rains, or in spring when the snow melts, the water in the channel must be rich in human and animal wastes. The author has seen no data of the water quality of the channel, nor a hydraulic tracer study on how the channel discharges influence the intake of the surface water treatment plant.

Repair work was done on the coagulation, flocculation, and sedimentation processes during the spring of 2001 and it seem that they had difficulties restarting the processes.

The judicial inquiry (Laing[5]) focuses on this fact. This may have been the immediate cause of the outbreak. But there are larger problems. To place a water treatment plant 2 km downstream from the cloaca maxima of North Battleford seem to be disingenuous and in gross violation of all public health considerations. The very fact that the plant had to use potassium permanganate (McDonald[6]) for taste and odor control indicates that the plant was taking in water laden with sewage, storm water, and industrial wastes. The use of water from this location is possible only with a much more advanced technology. (The space station recycles waste water after treatment with reverse osmosis)

The obvious solution to the problem is a modern groundwater system upstream of the city, and closure of the downstream surface water plant. For the upstream system one might consider a lagoon with about 7 days of storage in case upstream agricultural uses cause peaks of pollutant loads.

The interesting fact of the case is that the gastrointestinal cases increased the sale of antidiarrheal drugs and that it was noticed. The disease peaked in April, and the "boil water" order was in effect for 13 weeks, or 3 months after the beginning. What is also interesting that the concentration of *Cryptosporidium* in the waste water effluent was roughly 10,000 per liter in May indicating that there were quite a number of active cases in town. The finger pointing at Edmonton, 250 km upstream, is besides the point, but perhaps cattle operations upstream of North Battleford swept manure into the river upstream of the city.

The surface water plant had been in operation since 1950 serving initially mostly a mental hospital. It may be instructive to look at cases of gastrointestinal illness of these patients which may show the occurrence of the "spring flu". Eskimos have a saying: "Don't eat the yellow snow". North Battleford residents should be required to memorize and recite this, and take the appropriate actions.

2.3   The Walkerton, Ontario, *E. coli O157* outbreak

Walkerton is a small town in farm country in the Canadian Province of Ontario about 150 km west of Toronto. It has about 5000 inhabitants. Its water supply comes from two deep wells, and one shallow well. The deep wells are about 70 meters deep reaching into fractured limestone and produce very hard water. The shallow well Nr. 5 is 15 meters deep, was built in 1978, and was close to a barnyard and a swampy area. When pumping started, nearby springs stopped flowing. The water from this well was much lower in hardness, and the resident's liked it for its lower hardness. Practically every house had also a cistern to catch the rainwater, which the locals preferred for washing. The cisterns had pumps that would pump the water into the water supply lines of the house. Backflow preventer valves were supposed to prevent the cistern water from entering the distribution system of the town.

On May 12, 2000, large rain storms swept through Ontario, and dumped over 10 cm of water in a matter of hours. The past few years had been very dry, and the rains made quite a mess and everybody was involved in clean-up activities. Soon thereafter almost the entire population of Walkerton came down with excruciating stomach cramps and bloody diarrhea. The local hospital was overrun with patients, and reported it to the provincial health offices. Initial inquiries to the local utility regarding the drinking water were negative, and a food-borne source was expected. The health department, nevertheless, issued a "boil water order". After more tests it was discovered that the water contained high numbers of *E. coli*. More detailed tests revealed that the water and specimens of stool from the sick people contained the subgroup *E. coli* O157:H7, which causes the sometime fatal HUS (hemolytic uremic syndrome) Many patients, especially children, developed serious kidney problems, and were flown to hospitals in other cities. At the height of the crisis, over 2000 people were ill, and 7 died. Bottled water was brought in, and the local hospital had a water supply system taking water from milk trucks. What went wrong?

After a thorough investigation, and after great public pressure by the media, and the usual lawsuits which are still going on, the prime target of the *E. coli* incursion into the supply system narrowed to the shallow well Nr. 5. This well is on the outskirts of the city on the southeast side in rolling farmland. Further to the east, about 200m from the well, and on elevated land, the local veterinarian kept a herd of about 100 cows. He practiced the normal and proper drain management practices and kept the manure at a concrete bin.

The extremely heavy rain, however, must have flushed some of the manure down the hill towards the well, where it infiltrated and reached the well in a matter of hours.

The chlorinators, according to the newspapers, worked only sporadically, which is probably a euphemism for not working at all. A check on chlorine consumption and orders showed a rather low use, and that a proper chlorine residual was not maintained. Some of the reports on concentrations were not based on actual samples.

On a visit to Walkerton in August, 2000, water samples were taken at the cemetery, the public library, the only sports bar in town, and at the motel where the author stayed. All samples showed a chlorine residual in excess of 2.0 mg/l free chlorine. The total plate count (HPC) was under 50CFU/ml for all samples; in other words the water was safe. But the boil-water order was still in effect.

There were other strange happenings. All the shower heads in the houses had been replaced by new ones on the suspicion that they might harbor the *E. coli*. Simply boiling the heads would have been sufficient. Free bottled water was still available at a warehouse in town. All fire hydrants were secured with steel bands to prevent an opening. It was on a Friday evening, and the crews of plumbers brought into town were celebrating and ready to leave town. They had gone house to house and inspected all the plumbing, removed the backflow valve connectors in the houses, and made the cistern systems completely separate from the public water supply system. Rumors were that some citizens objected to this. Their public water supply connections were shut off. In a panic situation many things are done, and many do not make any sense. Why the ministry of health waited to lift the boil-water order until December is not clear.

In simple technical terms the problem was: A shallow well under the influence of surface water, was inundated with rainwater runoff from a feedlot. Figure 1 below shows a sketch of the situation that is not to scale.

Figure 1 *Cross section profile of Walkerton, Canada*

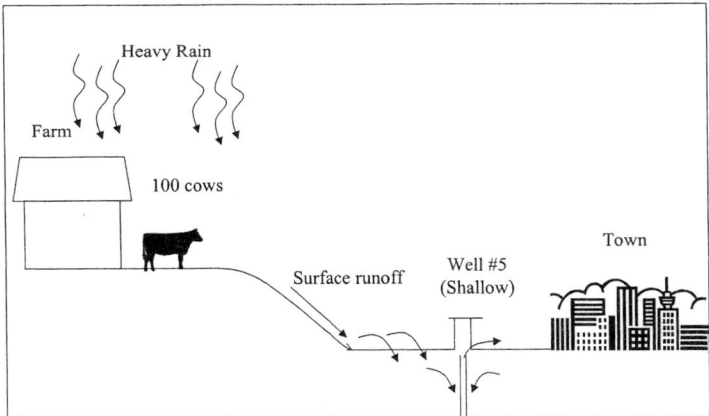

The chlorinators were not functioning, and *E. coli* laden water was sent through the system for several days. The whole thing probably happened before, but not in this magnitude and probably only a few people became ill.

2.4   The Gideon, Missouri, Salmonella outbreak

Gideon, Missouri is a small town with a population of about 1,100 people. It is in rural Missouri and is served by ground water wells. In late November of 1993 a number of young and old people became seriously ill and were hospitalized. Salmonella was identified as the causative agent, and at first a food-borne source was expected. However the patients had not shared a common meal or had been at the same social function. So the most likely cause was the water supply. Subsequent water samples were positive for fecal coliforms. On December 18 a "boil water order" was issued. The epidemic ended in the middle of January 1994 and it was estimated that about 500 people became ill, and seven residents of a nursing home died.

What happened?

In the middle of November the town had received many complaints of a taste and odor problem and started a flushing program through their 50 hydrants. Reports from field visits by EPA showed that the well water was not contaminated. The town has two water towers with a capacity of 100,000 and 50,000 gal respectively. There was also a private 100,000 gal water tank that was offline and had been drained. The large tank was partly covered and had open access covers. A large number of pigeons roosted on the tank, and bird droppings fell into the tank. The droppings must have accumulated in the tank sediments. During flushing of the system these sediments were washed into the system and modeling (Rossman[8]) showed that the houses with affected people were in the direct neighborhood of the tanks. The general conclusions are that the tanks caused the problem.

2.5   Detroit Main Break, no outbreak

On Thursday, June 10, 1999 the Detroit News carried on its front page the story of a break of a water main and it's consequences. There were pictures of people picking up bottled water, and more ominously a picture of surgeons at a nearby Hospital scrubbing down before an operation with sterile water from water bottles. What happened the day before is that a crew laying a fiber optic cable had punctured and fractured a 42 inch water main that was bringing water to the northern suburbs of Detroit from a treatment plant at Lake Huron. The pipe was isolated, and in the process some customers lost water, and others had water at low pressures. A large GM assembly plant, the headquarters of Daimler –Chrysler, several area schools, and a 200 store shopping center had to close down. All in all, over 30,000 employees had to be sent home in Oakland County. People were advised not to drink the water at the homes that had water service, and those that had none were told to go to fire stations where bottled water was available from a variety of corporate sponsors.

It took several days to fix the main, and the boil water order was eventually lifted. Economic losses were estimated at $7.5 million per day. The newspapers, radio and TV stations carried the warning messages:

Obtain bottled water for cooking and drinking

Treat tap water used for cooking and drinking by boiling it for 5 minutes and cooling it for 5 minutes

Disinfect water by adding 3 drops of household bleach to every gallon of water. Stir and let it sit for 30 minutes.

Water does not have to be boiled for hand washing and bathing.

We saw the pictures in the newspaper, and we called the hospital and told them that we would like to take water samples. They reluctantly agreed, and we drove 90 minutes to the hospital. When we arrived they were relieved that we did not want to sample at the operating rooms, but just the drinking fountains in the lobby. The fountains were operating, but had large signs on them not to drink the water. We sampled the two fountains and told the receptionist that the water was fine (less than 50 cfu./ml). On our way home we could not resist to pick up a gallon of bottled water at the closest fire station. We stopped at the local McDonald's to get a cup of coffee. The sign on the door read: "We have no water". So there was no coffee, only bottled drinks. Off course they had water. The toilets functioned and we took a sample. Not from the bowl, but from the tank. Upon return to the lab the water turned out to be fine; the only high count we found was in the bottled water (400 cfu/ml) We emailed the director of the hospital laboratory the same day that his water was fine. He replied that he expected this, but it was " Better to be safe, than be sued". And indeed, a few days later we read in the papers that an enterprising lawyer sued the city not for the health consequences of drinking the water, but for his mental anguish that he might have drank contaminated water because he found out about the incident too late.

3 LESSONS LEARNED

Outbreaks are bound to happen, but hopefully lessons have been learned. The prime detection systems at the moment are the emergency rooms of hospitals. This means that it is a bit late, and people are ill. The earlier an outbreak is detected, the faster a response is possible. A reporting system on the absenteeism at schools, increased gastrointestinal illnesses at nursing homes and day care centers are an indication that something is going wrong in the community. Senior citizens serve a useful purpose like canaries in mines.

All of the above is not very satisfactory, and we badly need more responsive detection systems that report the water quality in near real time. With regard to protozoa, a particle counting system can be online, and may detect a breakthrough of particles early enough.

Bacteriological methods are simply too slow at the moment. Continuous monitoring of bacteria is possible by looking for their ATP (Deininger[3]), but the system is expensive and out of reach for many utilities.

So the best surrogate parameter to monitor is the disinfectant residual. Monitors for disinfectant residuals exist and a placement of them at key points in the system and a continuous display of the status in real time is the best protection.

If an event happens, the prime consideration has to be a public notification program via loudspeakers, radio and television, and the standard boil water orders must go out as rapidly as possible.

The clean-up and flushing of the system is not a trivial undertaking and should be exercised using harmless tracers.

References
1   R.M. Clark et al., *Journal of Water Supply Research and Technology - Aqua*, 1996, 45 (4), pp. 171-183.
2   R.M. Clark. and R.A. Deininger, *Journal of Contingencies and Crisis Management*, 2000, 8 (2), pp. 73-80.

3   R.A. Deininger and J.Y. Lee, *Field Analytical Chemistry and Technology,* 2001, **5**, (4), pp. 185-189.

4   K.R. Fox and D.A. Lytle, Milwaukee's Crypto Outbreak: Investigation and Recommendations, 88 (9), September 1996, pp. 87-94.

5   R.D. Laing, *Opflow, 2002,* 28 (6), June 2002, p.12, See also *www.northbattlefordwaterinquiry.ca*

6   R.A. McDonald, *Proceedings, Am. Water Works Assoc. Annual Conference*, New Orleans, June 2002

7   D.R. O'Connor, *Opflow, 2002,*28(6), p. 8, June 2002. Further details can be found at *www.walkertoninquiry.com*

8   L.A. Rossman, R.M. Clark and W.M. Grayman, *J. of Environmental Engineering*, 1996, ASCE, 120(4), pp. 803-820.

RADIOLOGICAL ISSUES

O. D. Hydes
Independent Drinking Water Consultant

1 INTRODUCTION

Historically there has been little interest in radioactivity in water supplies. But that changed in 1986 with the Chernobyl nuclear accident when it was realised that the Government, water suppliers and other relevant organisations were not adequately prepared to deal with potential major radiological contamination of water sources and water supplies. This prompted the Government to issue[1] general advice on radiological incidents and emergencies affecting the environment, including the role of Government Departments, water suppliers, local authorities and other relevant organisations. This was followed by a supplement[2] that gave detailed advice to water suppliers on sampling and analysis of public water supplies following the accidental release of radioactivity to the environment.

2 ROUTINE MONITORING OF WATER SOURCES AND WATER SUPPLIES FOR RADIOACTIVITY

Currently water suppliers do not routinely monitor water supplies for radioactivity because there are no specific requirements to do so under current legislation. However under the provisions of the Euratom Treaty, the Environment Agency (EA) and its predecessors, on behalf of the Department for the Environment, Food and Rural Affairs (DEFRA) and its predecessors, manages a routine monitoring programme for radioactivity in water sources. This monitoring started many years ago and originally included about 30 water sources (rivers, reservoirs and ground waters) selected because of their proximity to activities involving the use of radioactive substances. There have been a few changes in water sources over the years and now a few treated water supplies are included. It is estimated that the water sources and supplies monitored cover the water supplies to between 10 and 15% of the population of England and Wales. A small sample is taken by the water supplier from the sampling site each week and bulked to give a three monthly sample that is then analysed by a contractor laboratory to the EA. Each site is monitored for gross alpha and gross beta activities and for a range of individual radionuclides. The results of this monitoring are published annually by DEFRA.[3]

This monitoring has shown that one of the ground water sources regularly exceeds the World Health Organisation (WHO)[4] screening value of 0.1 Bq/l for gross alpha activity. This source is blended with other sources so that the gross alpha activity in the water supply is less than 0.1 Bq/l. Two other sites occasionally exceed 0.1 Bq/l for gross alpha activity. None of the sites monitored exceed the WHO screening value of 1.0 Bq/l for gross beta activity.

3   THE EC DRINKING WATER DIRECTIVE

The new drinking water Directive,[5] which comes fully into force on 25 December 2003, contains some requirements in respect of radioactivity. It does not set any mandatory standards for radioactivity in drinking water but it does include non-mandatory indicator parameter values of 100 Bq/l for tritium and 0.1 mSv/a for total indicative dose. Total indicative dose does not include tritium, potassium-40, radon and radon decay products. An indicator parameter value is set primarily for monitoring purposes. When an indicator parameter value is exceeded, the water supplier must investigate to determine the cause and whether the exceedence is likely to recur. The water supplier will only be required to take action if the extent and duration of the exceedence is a risk to human health.

The Directive does not set out specific monitoring requirements for the two radioactivity indicator parameters. It states that the monitoring frequencies, monitoring methods and most appropriate locations for monitoring points will be decided in accordance with the procedure laid down in Article 12 of the Directive. This means that proposals from the European Commission are considered by a Committee composed of representatives of the Member States and chaired by the Commission. Draft proposals have been put to the Committee but agreement had not been reached at the time of writing this paper.

The draft proposals consist of monitoring at the audit monitoring frequencies in Annex II with a staged monitoring procedure for total indicative dose. The first stage proposed is monitoring for gross alpha and gross beta activities. If the WHO screening values were met no further action would be required. If either of the WHO screening values is exceeded, the water supplier would be required to carry out specific radionuclide analysis, the radionuclides to be sought depending on the natural and artificial sources of radioactivity in the catchment, and to calculate whether the total indicative dose is met from a formula based on the observed and reference activities of the specific radionuclides. There is also a provision in the Directive whereby if it can be demonstrated that the tritium and total indicative dose values are well below the indicator parameter values, monitoring for radioactivity need not be carried out.

4   THE DRINKING WATER QUALITY REGULATIONS

The requirements of the Directive have been transposed into law by the Water Supply (Water Quality) Regulations 2000[6] that apply to the "English" water suppliers and the Water Supply (Water Quality) Regulations 2001[7] that apply to the "Welsh" water suppliers. Similar regulations have been made in Scotland[8] and Northern Ireland.[9] These Regulations include the provisions for radioactivity in the Directive and they anticipate the outcome of the Article 12 Committee by requiring water suppliers to monitor for tritium and gross alpha and gross beta activities from 1 January 2004. There is a provision in the Regulations for a legal notice to be issued to water suppliers for additional or different monitoring for radioactivity should the outcome of the Article 12 Committee be different from that anticipated.

Any exceedence of the values for tritium and total indicative dose will be investigated by the water supplier. It will be the Drinking Water Inspectorate (DWI) in respect of England and Wales, in consultation with its medical advisers that will make the judgement as to whether the extent and duration of any exceedence is a risk to human health. If it is the water supplier will be required to taken action such as treatment (including blending) to reduce the concentration of radioactivity in the supply or negotiation with the EA to reduce discharge of radioactivity to the water source. Information available suggests that all water supplies in England and Wales will be well below the tritium value of 100 Bq/l and below the gross beta screening value of 1.0 Bq/l. However a few water supplies will exceed regularly or occasionally the gross alpha value of 0.1 Bq/l but it is not known whether they will exceed the total indicative dose until the appropriate specific radionuclide analysis is carried out.

5   INCIDENTS AND EMERGENCIES INVOLVING RADIOACTIVITY

The responsibilities of Government Departments, water suppliers, local authorities and other relevant organisations are set out in a 1995 publication of the then Department of the Environment, popularly referred to as the "Green Book".[10] This publication is a revision of the guidance issued following the Chernobyl accident. It includes advice to water suppliers on sampling and analysis. It is understood that DEFRA has commissioned a revision of the "Green Book" because there have been numerous changes in the responsibilities, structures and names of Government Departments and all the other relevant organisations involved.

The "Green Book" covers several types of incidents and emergencies. DEFRA is the lead Government Department for overseas nuclear accidents and it would implement a national response plan involving action by a number of Government Departments, the EA, water suppliers, local authorities and others. The plan is supported by the Radioactive Incident Monitoring Network (RIMNET). As part of the plan the EA would set up a Technical Co-ordination Centre that would provide advice to DEFRA and DWI and the latter would advise water suppliers. Department of Trade and Industry (DTI) is the lead Government Department for domestic nuclear accidents and would set up the Nuclear Emergency Briefing Room. The operator of the facility where the accident occurred would set up an Off-site Centre that would include EA and the local water suppliers. DEFRA would set up an Environment Operations Centre that would include DWI who would advise water suppliers. The "Green Book" also covers the arrangements for other types of radiological accidents including transport of radioactive material, satellites and defence installations.

All water suppliers are required to prepare an emergency plan by the Security and Emergency Measures Direction 1998.[11] This plan has to set out how the water supplier will deal with any emergency affecting the sufficiency and quality of water supplies and this would include any radiological emergency. The plan has to include the nomination of a person to be responsible for managing all aspects of the response, the liaison with DWI and other organisations, the sampling and analysis of water sources and supplies, the acquisition of toxicological advice, decisions on intervention (issue of advice not to use the water and provision of alternative supplies) and liaison with, and provision of advice to, consumers and the media. The Direction and the associated guidance also requires water suppliers to make available within 24 hours, an alternative supply of at least 10 litres per person in the event of an unavoidable failure of supply and to continue the alternative supply of at least 10 litres per person each 24 hours until the normal supply is restored. Priority for alternative supplies should be given to vulnerable groups such as hospitals, schools and bottle-fed babies.

6   ISSUES RELATING TO RADIOLOGICAL EMERGENCIES

6.1 Sampling and analysis

The "Green Book" provides advice on sampling and analysis of water sources and supplies in the event of a radiological emergency. Sampling and analysis should be carried out as quickly as practical and should be agreed with EA, DWI and DEFRA. For an overseas nuclear accident these organisations would issue advice to water suppliers. For a domestic nuclear accident initially surface water sources would be sampled within a 30^0 sector downwind to 16km distance and extended as necessary. The "Green Book" advises that measurement of gross alpha and gross beta activities is not appropriate in an emergency because information is needed about the specific radionuclides present in order to decide whether intervention is necessary. RIMNET only accepts results if the quality assurance of the results is satisfactory.

Water suppliers have limited analysis facilities for radiological examination of water sources and supplies. Only four have the special analysis equipment needed. This is sufficient to cope with normal routine compliance analysis and, through the water industry mutual aid scheme, a nuclear accident that affects a limited area of the country. It would not be sufficient to cope with an accident that affected the whole country or a substantial part of it. In such circumstances the water industry would seek to use external analysis facilities but in the event of a major nuclear accident there would be considerable pressures on these facilities for analysis of other media such as air and food. The author considers that there should be a review of whether there are sufficient radiological analysis facilities for all media, including water, to cope should there be a major nuclear accident.

Pressure on water analysis facilities could be eased if it was acceptable to carry out initial screening for gross alpha and gross beta activities. Such screening could indicate whether there has been any significant radiological contamination of water sources and supplies. If contamination is indicated then specific radionuclide analysis should be carried out. The author considers that the revision of the "Green Book" should reassess the value of screening for gross alpha and gross beta activities in the event of a nuclear accident. The author also considers that DEFRA should allow results of water analysis to be accepted by RIMNET with a qualifying comment when the analysis does not meet the quality assurance requirements because in an emergency all information is useful provided its limitations are understood.

6.2 Toxicological advice and intervention

Toxicological advice and advice on intervention is given in the National Radiological Protection Board's (NRPB) guidance on the restriction on food and water following a radiological accident.[12] This recommends the use of Council Food Intervention Levels (CFILs) for liquid foods from the European Commission as the action levels for restriction on the use of water supplies for drinking and cooking. These CFILs represent a balance between the possible harm from the dose of radioactivity and problems posed by applying restrictions to water supplies. They could be revised following an accident. NRPB emphasises that water may be consumed safely for short periods at levels well in excess of the CFILs and it may be used for longer periods for washing and toilet flushing at even higher levels. NRPB states that immediate withdrawal of drinking water supplies is not generally necessary, but that every effort should be made to provide alternative supplies quickly in order to maximise the dose reduction achieved. The following table gives the CFILs for liquid foods and water.

Table 1 *Council Food Intervention Levels for liquid foods and water*

Radionuclide Group	CFIL (Bq/l) for liquid food/water (1)
Isotopes of strontium, notably ^{90}Sr	125
Isotopes of iodine, notably ^{131}I	500
Alpha emitting isotopes of plutonium and transplutonium elements	20
All other radionuclides of half-life greater that 10 days, notably ^{134}Cs and ^{137}Cs (2)	1000

(1) The sum of the detected concentrations of all radionuclides in the group
(2) This group does not include ^{14}C, tritium and ^{40}K.

The author accepts that the use of CFILs are a sound basis for deciding on intervention, but notes that the levels are orders of magnitude greater than the WHO screening values for gross alpha and gross beta activities for normal supplies. Also the advice on how quickly intervention should be made when a CFIL is exceeded is somewhat vague. The author also considers that public perception needs to be taken into account when considering intervention. If there was a major radiological accident that affected water supplies it is likely that media reaction and public perception would demand intervention at levels below the CFILs. The author considers that the relevant organisations should review the action levels at which intervention should be considered for water supplies.

7 CONCLUSIONS

Major radiological accidents are difficult to deal with. There are many players involved, Government Departments and other organisations, and good communication and co-ordination is needed. The "Green Book" needs urgent revision in view of all the changes in organisations and their responsibilities. The author is not aware of any recent rehearsal of emergency procedures to deal with a radiological accident and recommends that such a rehearsal is undertaken soon after the revision of the "Green Book" is complete. This rehearsal should test the ability of the water suppliers and other external analysis facilities to cope with the demand for radiological analysis of all media, including water sources and supplies. The relevant organisations should review the action levels for intervention and the provision of alternative water supplies in view of the likely public perception of a radiological accident.

References
1   Department of the Environment, Guidance for Dealing with Incidents and Emergencies Involving the Release of Radioactivity to the Environment, 1987.

2   Department of the Environment, Supplement No 1 to Guidance for Dealing with Incidents and Emergencies Involving the Release of Radioactivity to the Environment, 1988.

3   Department for Environment, Food and Rural Affairs, Digest of Environmental Statistics, published annually.

4   World Health Organisation, Guidelines for Drinking Water Quality, Volume 1 Recommendations, 1993.

5   Official Journal of the European Communities, Council Directive 98/83/EC of 3 November 1998 on the quality of water intended for human consumption, L330, Volume 41, 5 December 1998.

6   The Water Supply (Water Quality) Regulations 2000, SI 2000 No 3184, The Stationery Office Ltd, 2000.

7   The Water Supply (Water Quality) Regulations 2001, The National Assembly for Wales, SI 2001 No 3911 (W.323), The Stationery Office Ltd, 2001.

8   The Water Supply (Water Quality) (Scotland) Regulations 2001, Scottish SI 2001 No 207, The Stationery Office Ltd, 2001.

9   The Water Supply (Water Quality) (Northern Ireland) Regulations 2002, Statutory Rules of Northern Ireland 2002 No 331, The Stationery Office Ltd, 2002.

10 Department of the Environment, Civil Emergencies Involving Radioactive Substances: The Department's Role and Arrangements, June 1995.

11 The Department of the Environment, Transport and the Regions, The Security and Emergency Measures (Water and Sewerage Undertakers) Direction 1998.

12 National Radiological Protection Board, Documents of the NRPB, Guidance on Restriction on Food and Water following a Radiological Accident, Volume 5, No 1, 1994.

COMMUNICATIONS

C. B. BUCKLEY

"Precision of communication is important, more important than ever, in our era of hair-trigger balances, when a false, or misunderstood word may create as much disaster as a sudden thoughtless act."[1]

1 INTRODUCTION
The significance of precise, well delivered and equally well understood communications, together with the effective use of well defined and user friendly communications pathways can never be over exaggerated nor underestimated in establishing and maintaining confidence in the management, investigation and resolution of emergencies, particularly those involving the contamination of public drinking water supplies.

Despite improvements in communication systems during the last few decades of the 20th century, failures to liaise and communicate effectively, thus contributing to the cause and development of water quality incidents remain many and varied. Reasons for these findings can be proffered but can they continue to be justified and acceptable to a more discerning public. Lessons from previous calamities must be learnt. Securely structured systems must be implemented that must also be well rehearsed, to obviate uncertainty but without resulting in over familiarity.

To emphasise these crucial aspects, I have selected three well-known water contamination incidents from the 1980s, the decade before the structure and organisation of the water industry changed for the second time in twenty-five years. These demonstrate that some faults and inadequacies change little with the passage of time, even though the water industry is now so well regulated, with improved systems of work emerging from the widespread use of Quality Assurance. But are communications Quality Assured?

2 THE BRAMHAM INCIDENT, 1980

A severe outbreak of gastro-enteritis in the Bramham[2] rural area of Yorkshire, which is situated North East of Leeds, in late July 1980 was rapidly attributed to microbiological contamination of the public water supply. A study carried out by the local Medical Officer of Health on 1,000 of the estimated 5,000 households revealed that at least one member of each household had experienced gastro-enteric symptoms before the water utility had issued an advisory boil water notice to the community. This meant that almost 3,000 of the estimated 12,000 population had been affected. Faecal specimens from 43 patients were examined but no specific causative bacterial or viral causal agent was identified, neither were there any cases of hepatitis.

The sequence of events and circumstances that preceded this outbreak highlighted lack of effective communication as being of major significance. The source of the potable water supply to Bramham[3] was local, comprising four 12" diameter boreholes in Magnesian Limestone and varying in depth from 31m to 38m. The first borehole had been developed in 1937 followed by Boreholes 2 and 3 in 1965. Borehole number 4 was sunk in 1974 but not commissioned until 1979. This borehole was the nearest to the Carn Beck stream that runs through the village, the stream receiving storm sewage overflows from the local sewerage infrastructure.

The abstracted source water was softened and disinfected at Bramham Pumping Station, with maintenance of the chlorination unit being carried out at six monthly intervals under contract, by Wallace and Tiernan, the suppliers of the chlorination equipment. This equipment had been serviced in February 1980 some five (5) months earlier, when Wallace and Tiernan (W & T) had immediately sent to the utility, replacement parts for those that had shown signs of wear during their February visit. W & T had offered to visit the site to replace these parts, prior to their next scheduled service in August 1980, but were told that this was not necessary.

Samples to check the microbiological quality of both the raw and treated water at Bramham were taken on a weekly basis, with re-samples being scheduled if positive results for coliform organisms were observed.

Staff collecting the samples from the Bramham site were not aware of the boreholes in use at the time of sampling. It was customary to pump only two boreholes on a daily basis, with the combination of borehole being rotated. Consequently any re-samples that were collected following the detection of coliform organisms in the original sample, could be from a different combination of boreholes.

From late June, a series of samples had 'failed' i.e. coliform organisms had been detected, in both raw and treated waters. Although these results had been reported to staff responsible for operating the boreholes, during mid July, 'failures' were attributed not to unsatisfactory water quality but to the possibility that non-sterile sampling bottles had been inadvertently used. Checks to confirm that the autoclave used to sterilise glassware and media was working effectively were inconclusive, so microbiologically positive results for many samples were largely dismissed. Arrangements to re-sample affected sites were delayed due to staff shortages (holidays) and the intervening weekends.

Residual chlorine measurements were normally taken by staff visiting the site for water quality sampling purposes. Due to the close proximity of customers to the pumping station, residual total chlorine values were normally kept at or below 0.01mgl^{-1}. Continuous recording chlorine meters were not installed at this site. But operatives estimated chlorine levels by the daily depletion of the sodium hypochlorite tank. When it was realised that there was a possibility that the water supply was contaminated, the chlorination unit was examined and found to be faulty - a failed carbon ring had allowed the sodium hypochlorite solution to be discharged to waste, passing into the gland lubricating water. This meant that the chlorination unit only appeared to be functioning correctly.

Coincidentally, during the period when samples of raw and untreated water contained coliform organisms, and the chlorination equipment was faulty, the pollution prevention department of the water utility informed Leeds City Council's Main Drainage Department that the stream in Bramham was polluted. A few days later, the waste water department of the water utility independently contacted the Leeds City Council's Mains Drainage Department, who were the Sewerage Agents for the water utility, requesting that they carry out an inspection of the sewerage system in Bramham village, to check for possible pollution of the local boreholes.

A blocked sewer was discovered. This had caused the storm sewage overflow to operate, polluting the stream with sewage debris. The sewer was unblocked immediately and the stream receiving the sewage overflow was cleaned the next day.

This was a normal course of action carried out by sewerage agents, but the significance of the sewage discharge was not wholly appreciated due to

(i) the sewerage agents were not aware of any potable water sources being in close proximity to the sewerage system, and

(ii) the potable water quality department of the utility were unaware of the discharging sewer and consequential pollution to the stream.

The CCTV inspection of the sewer in Bramham also revealed several badly constructed and faulty connections where leakage into the strata could occur.

It is well established that Magnesian Limestone aquifers rely upon fissures for the transfer of water. These are subject to hydraulic short-circuiting of water and of course, ingress of contaminants. A Report on the boreholes as sources of supply had been prepared by the water utility in the mid 1970s. The Director of Operations of the water utility prepared a Report following his Inquiry into the Contamination of the Boreholes at Bramham during July 1980, commenting that an earlier internal document on these boreholes had not made reference to either water quality nor of the effect of pumping the boreholes, on the stream.

Not surprisingly, this borehole source was quickly abandoned and the many recommendations of the Director of Operations were considered and acted upon. Several of them related indirectly and directly to glaring gaps in Communication pathways both internally and with external organisations!

3 POLLUTION OF THE RIVER DEE - JANUARY 1984

In late January 1984, potable water supplies to circa two million domestic, commercial and industrial customers in parts of North East Wales, Merseyside and Cheshire were contaminated with chlor-phenolic compounds. Almost ten days elapsed before all traces of these highly objectionable tasting contaminants could be flushed from the treatment works and the supply and distribution networks of the four water utilities - Welsh Water, North West Water and the former Chester Water Company and the Wrexham and East Denbighshire Water Company (these now form Dee Valley Water), that were abstracting water from the River Dee.

Although intensive abstraction from the River Dee as a source of potable water supplies began in the 1950s, there was no previous records of any major pollution incidents affecting the integrity of potable water supplies. There were however, about fifty (50) taste and odour related incidents annually with several of these being linked with snow melt conditions in the catchment area that increased surface water run off into the River Dee and its tributaries. There had been a sudden thaw a few days prior to this incident.

The cause of this incident was investigated, together with a review of the management of the incident and the subsequent clean-up operation of the treatment, supply and distribution infrastructure. This led to a formal document - 'Joint Report on the Pollution Incident affecting the River Dee and Water Supplies, January 1984'[4] being submitted to the then Department of the Environment in London and the Welsh Office in Cardiff. This Report emphasised the urgent need for a Working Group to be formed immediately to consider and report upon -

(i) Intercommunication arrangements between the four potable water abstractors;

(ii) Communication with Government departments, local authorities, health authorities, major agricultural, commercial and industrial customers;

(iii) Improving arrangements for assessing the severity of pollution incidents and any consequential risk to public health.

A specific task group[5] considered communications and identified a significant number of deficiencies in the communication systems existing at the time of this incident. During the incident, all abstractors experienced major difficulties in dealing with the huge numbers of complaints that besieged and totally blocked the water utility and local authority switchboards with incoming calls. With only very few direct and/or ex-directory telephone lines available to the water utilities to make contact with –

- field staff
- other involved water utilities
- local authorities
- health authorities
- media
- major customers

it was hardly surprising that investigatory and remedial work was hindered. Quite obviously, this difficult situation contributed to the queries and complaints that were received. Also the level of criticism aimed at each water utility by the media and the public.

Not surprisingly, the following deficiencies were highlighted: -

(i) A lack of formalised procedures for communications between water utilities abstracting water for potable supplies from the same river system.

(ii) No means of adequately communicating data and information between abstractors other than by telephone.

(iii)No established procedures for implementing joint emergency control to co-ordinate activities, should a major pollution incident occur.

(iv) No formal guidelines for categorising problems linked with deteriorating river water quality.

(v)Lack of uniform procedures for swift communications with Environmental Health Officers, Medical Officers of Environmental Health (now known as Consultants in Communicable Disease Contact (CsCDC) about deleterious changes to the quality of potable water supplies.

(vi) No guidelines on the mechanism for alerting customers and the media to adverse changes in the quality of potable water supplies and risks to public health.

(vii)Lack of procedures for contacting Government departments during such situations.

These deficiencies were rapidly addressed with recommendations for significant improvements in communications being accepted and progressively introduced during the six month period up to September 1984.

Many of these beneficial changes have stood the 'test of time', being subject to only minor amendments, as the water industry changed its structure from ten all-encompassing water authorities with responsibility for potable water supplies; waste water collection and treatment, pollution control and environmental protection, to their new status when the water utilities were privatised in 1989 and responsibility for pollution control and environmental protection was relinquished, initially to the National Rivers Authority and more recently the Environment Agency.

A series of formal situation reports were devised and implemented to inform each of the four abstractors of relevant river water quality and conditions. These were issued daily at noon and were known as DEESIT Reports.

The foregoing was supplemented by a system of categorised alarm states that described the status of polluted river water. There were three such categories -

Category 1 -   Major incident requiring operational action at treatment sites and downstream abstractors and co-ordinated external publicity

Category 2 -   A potential serious hazard to river water quality, the extent of which is being actively investigated.

Category 3 -   Minor pollution suspected, the situation being maintained.

In order to emphasise the importance of communications relating to these alarm states, the prefix of DEEPOL was used. The current descriptions of these alarm states as used by the Environment Agency remain substantially the same, but are not surprisingly more detailed.

The Communications Task Group also made a number of far reaching recommendations for change,

(i)   Introducing the use of telefacsimile equipment and electronic mail to help ensure speedy, consistent contact with public health specialists and other functional advisers.

(ii) Initiating a series of regular review meetings between the utilities, public health organisations and Environmental Health Officers on topics of mutual concern and interest.

(iii) Formalising procedures for issuing statements to the media and to customers during incidents or emergency situations.

The impact on the water industry of the contamination of River Dee - derived potable supplies was widespread and led to a Report on Actions to Minimise the Effect of Pollution Incidents affecting River Intakes for Public Water Supplies being issued by the Water Authorities and Water Companies Associations in June 1984. This Report[6] was produced by a group comprised of senior managers in the Water Industry, and senior Government officials, chaired by W. F. Ridley, Chief Executive of Northumbrian Water; becoming known as the 'Ridley Report'. Of the twenty recommendations in this Report, two were focussed upon communications and in particular, contingency plans for dealing with incidents, these to include consultation arrangements with external public health organisations, the public and the media. The 'Ridley Report' also highlighted the continuing need to ensure that there was a regular review of existing emergency procedures, particularly assessing the availability of communication resources such as telephones, radio pagers etc., and the appropriateness of details of utility personnel such as contact arrangements, especially out of normal working hours. These factors remain as being of continuing interest and concern!

4   THE LOWERMOOR (CAMELFORD) INCIDENT (1988)

During the late evening of 6 July 1988[7], South West Water Authority's Control Centre received a substantial number of complaints from customers in the Camelford area of North Cornwall, alleging that the public supply was dirty, discoloured and foul tasting with water based drinks becoming curdled. Subsequent complaints also alleged acidic, astringent tastes, with many expressing concern that following washing or bathing in the water, lips and eyelids were sticking together and body hair was sticking to the skin. Lips and mouths were also extremely dry.

Coincidentally, treatment problems - persistent malfunctioning of the lime dosing pumps - had been experienced throughout the day at Lowermoor Water Treatment Works (WTW), which supplies the Camelford area.

It was subsequently established that an unrecorded delivery of 20 tonnes of aluminium sulphate solution (8% Al_2O_3) had been inadvertently discharged into the locked control tank at Lowermoor WTW by a relief tanker driver. This action lead to widespread contamination of the water leaving the treatment works. Attempts to flush the contaminated water from the distribution network caused substantial environmental damage resulting in a major fish mortality and a major erosion of public confidence.

Dr. John Lawrence, who at that time was Director of the ICI Brixham Laboratory and a non-executive Director of South West Water Authority was asked to conduct an inquiry into the Incident. The Lawrence Report[8] was critical of the lax approach to operational security and the overall management of this incident particularly in its early stages. Communication with the public and the media at both local and regional levels was contradictory and confusing with the general statement regarding the cause of the incident being aluminium sulphate not being published until 22nd July. There were also shortcomings in the information regarding the location and supply of bowsers to customers during the period the distribution system was being flushed and cleaned.

Emergency Procedures had been issued by the Chief Executive upon his appointment, several months earlier in 1988, but these were not adhered to during the resolution of this Incident, consequently both internal and external liaison was initially at a low level with poor communication between local and Head Office staff, and between scientific and operational staff generally.

Prior to this Incident, the aluminium sulphate storage facility had been improved but not brought into use. The new tanks were located above the old concrete tank that was still in use, the top of which was partly open to atmosphere in the space between the new facility and the WTW building. The delivery of aluminium sulphate should have been made into this partly open, but unlabelled tank.

The delivery of aluminium sulphate was expected during the early part of the week of the Incident, so operatives were surprised to find the storage tank at a low level on the 8 July. A discussion with the supplier revealed that the delivery had been made on 6 July - two days earlier, when the customer complaints of unacceptable water quality started.

The regular tanker driver was unable to make this delivery, so he told the relief driver, who seemed to give the impression that he was familiar with the Lowermoor site, that the aluminium sulphate tank was 'on the left' after passing through the gates. This information was correct, but the contact tank manhole which was locked was also on the left of the access road on the site.

The regular tanker driver knew that the existing aluminium sulphate tank was not locked, so he only used the key to gain access to the site, if he arrived when it was unattended either during, or outside normal working hours. Having gained access to the site, the relief driver saw the locked cover of the contact tank alongside the access road, 'on the left', so when the key also opened this padlock he assumed he had located the correct storage tank.

Delivery drivers were asked to make telephone contact with a neighbouring works also on the edge of Bodmin Moor so that operations staff knew who was on site. But this action was not confirmed. Different telephone numbers were quoted on the order form and delivery note.

Coincidentally, the local Treatment Scientist had just changed his address, but had given his new telephone number to the utility's Control Centre for emergency use. When Control Centre staff attempted to call the Treatment Scientist to investigate the initial customer complaints, the person receiving the call gave the name of a restaurant, so Control Centre staff thought that they had been given, or were using the wrong telephone number. The Treatment Scientist had not mentioned that he had moved to live at the family's business premises. This code was aborted and attempts were then made to call other staff who were not on emergency stand by rota at the time. This led to confusion and delay in starting the investigation.

This Incident led to the residuary South West Water Authority being prosecuted at Exeter Crown Court in late 1991, being found guilty on the charge of Public Nuisance, with most of the allegations made by the Crown Prosecution Service being in the words of the Judge - "demolished".

Prior to the Crown Court case, a Lowermoor Incident Health Advisory Group chaired by Professor Dame Barbara Clayton had produced two Reports[9] commenting upon the implications for the health of the population in the Camelford area following the contamination of the drinking water supplies in July 1988. These Reports confirmed that the information and advice given to the local population was inappropriate and inadequate.

One of the major concerns to the public and to the police who subsequently investigated the Incident to obtain evidence for the prosecution was the apparent contradiction in expressing the strength of the aluminium sulphate solution. I had to provide a signed statement, explaining to the police the mode of expressing the strength of chemical solutions. This aspect had been raised by members of the public who claimed to have scientific knowledge. They asserted that the solution strength was circa 27%.

This Incident, yet again, confirmed that a series of unusual and unforeseen coincidences of events and human errors can combine together to create a lasting and highly damaging impact on public confidence due to the influence of a lack of effective and well-managed communications.

5 EMERGENCY SITUATIONS AND COMMUNICATIONS PATHWAYS

Management and control of emergency situations[10] must be separated from the normal day to day activities of an organisation, which, for example, could be either a water utility, an environmental regulator, or both. It is inevitable, nonetheless, that other business activities will continue in parallel to emergency situations. It is doubtful whether or not normal activities can be either reduced or temporarily suspended in such circumstances.

This means that dedicated emergency facilities must be available for immediate use at all times. Such facilities to include:

(i)   communication equipment;

(ii) correct procedures and operational records;

(iii) accurate lists of contact arrangements for staff and external organisations, various categories of customers, regulators and the media.

Experience should dictate that all staff involved in the management and communication aspects of emergency situations must be well prepared, with the self-discipline to maintain personal logs of their activities.

The Drinking Water Inspectorate[11] has helpfully provided a comprehensive, but not exhaustive list of the data and information required by them, to enable an effective assessment to be made of the way the situation is being handled. Much of this information should have been self evident to experienced water industry professionals. In concise terms the information required can be summarised as: -

(i)   What happened?

(ii) When did it happen?

(iii) How did it happen?

(iv) Who has been, or should be informed?

(v)   What are the risks?

(vi) What progress is being made to restore normal conditions?

(vii) Why did it happen?

(viii)Will it recur?

Prior to privatisation of the Regional Water Authorities in 1989, a document[12] was published in England and Wales, highlighting the importance of developing effective communication links between water utilities, local authorities, health authorities and public health experts, with all relevant personnel being aware of this crucial requirement. The main points of concern identified in Chapter 10 of this document also emerge from reviews of historical emergency situations within the industry. The three contamination incidents described in Sections 2, 3 and 4 of this paper provide ample evidence of recurring inadequacies and ineffective communications.

The need for utilities to ensure effective liaison between operational and scientific staff is emphasised in the 1988 edition of this document but this aspect is not included in a later edition[13] published in the 1990s. This, in my view is a serious omission, particularly in the current climate of the use of a multiplicity of contractors to carry out various tasks formerly undertaken by utility staff. Details regarding the exchange of data and information must be included in contractual requirements, in addition to incentive criteria for achieving performance levels.

Convening annual meetings between utilities, public health specialists, environmental health and other relevant organisations to discuss matters of mutual concern is essential. Badenoch[14] charged Health Authorities with the task of developing action plans to deal with outbreaks of cryptosporidiosis and giardiasis in communities.

Consequently, such meetings are now more generally acceptable than they were historically at the start of the initiative to build fruitful and reliable communications pathways between these important and influential contributors to matters of public health.

Chapter 7 of the Publication[15] that accompanied the Water Supply (Water Quality) Regulations 1989, in England and Wales, clearly states the action(s) to be taken by water utilities when water quality standards have been infringed, particularly if the infringement, or operational failure is of a magnitude that could prejudice or pose a threat to public health. The need to notify relevant local authority and health authority staff in accordance with regulatory requirements is emphasised.

However, experience[16] confirms that the final phrase - "without delay" - can be of concern to utilities since it is influenced by many factors directly beyond their control that can be encountered in establishing immediate contact with public health organisations and experts. These are primarily due to differences in the way that the working day and out of hours contact are managed in different organisations.

Organisations operate flexitime arrangements; some office switchboards are frequently closed until 0900 hours, at lunchtime and after 1700 hours. Recorded messages helpfully inform callers that "the office will open again shortly", or "leave your name and number and your call will be returned as soon as possible". During office hours difficulty can be experienced in contacting named persons, who may not be available for discussion for a multiplicity of reasons. Of course nowadays, there are likely to be less staff available to assist.

This is the age of communication technology, so theoretically almost everyone can be contacted, almost anywhere in the world at any time, but can they to be spoken to? Voicemail has become increasingly fashionable, especially in organisations that use direct line facilities. The recorded message frequently omits to inform the caller when the 'occupant' will return. Often times there is no arrangement to be re-directed to a 'live' telephone, to actually speak to someone.

Electronic mail is regarded as a major improvement to communication pathways and so it is normally, but not at all times. Messages sent in this way may not be read immediately, for a wide variety of reasons.

In Section 3 of this paper reference is made to the recommendation that emerged from the River Dee Communications Task Group viz., the use of DEESITS and DEEPOLS to inform water abstractors via telefax facilities of the various conditions existing in the River Dee, or an activity and/or during emergencies. This system has been further developed and is now widely used by the Environment Agency to notify downstream users and the pollution departments of local authorities of deleterious changes in water quality. The three categories of water quality as applied to emergency pollution incidents remain almost unchanged from those implemented following the River Dee pollution in 1984. Relevant information is issued by the Environment Agency[17] using a form known as a POLFAX, with the response pro-forma from the water abstractor being known as the WATPOL. This form must be used to acknowledge receipt of the POLFAX.

Likewise, if a water utility has an operational problem or is carrying out a water activity that is likely to pollute water resources, an operational note (OPNOT) giving relevant details of the situation is used to inform the Environment Agency. However, enquiries have revealed that the practice of using the WATPOL and OPNOT standard forms as convenient communication pathways, appears not to be practised in all regions of the Environment Agency.

6 COMMUNICATIONS - USE OF WORDS AND PHRASES

If misunderstandings and misleading or biased conclusions are to be avoided during an emergency situation, the content of messages being relayed must not be ambiguous. This is extremely important. It is essential to provide an identical message to all 'consultees' involved in the management, resolution and investigation of such situations, particularly when statements are issued to the media and the public. Depending upon circumstances, certainly at different times in an emergency situation, 'consultees' often interchange between being both 'donors and recipients' of messages.

There is, therefore, an obligation on all involved in such situations to be factual, accurate, to provide clarity of detail with due brevity, so that the message 'as given' and 'as received' has the same meaning and significance to both donor and recipient. There is, of course the well known example of a message relayed in the armed forces:

As Given: - send reinforcements we are going to advance.

As Received: - send three pounds four pence we are going to a dance.

Jargon, acronyms and unexplained abbreviations must be avoided at all times. I am aware that in certain contexts, MOBR means Management of Operational Business Risks. In other contexts, I am also aware that it can mean -

(i) Monitor Output, Balance Resources, or

(ii) Mains Operations, Bacteriological Risks.

Certainty of meaning, not interpretative licence is a necessity.

When drafting written communications and recording notes of discussions think carefully or ill-chosen words and phrases may return to "haunt" you! Your notes and statements must be thoroughly checked for accuracy and fact. Given a short period of respite, will you still understand the meaning and context of your notes?

Remember that what might be intended for internal confidential use may ultimately be issued externally, depending upon a wide range of circumstances. My Confidential Report on the Lowermoor Incident[7] was subsequently used by the prosecution in the case that went to the Crown Court!

Effective media communications are essential, due to their impact on customers and the media. Definitive announcements by the media are used by water utilities to inform the public of emergency situations. Care must be taken to help ensure that the targeted audience can be contacted, for example, the topography of Wales and the West Country has meant that reception from local radio and television outputs is poor or non-existent in some areas, but signals from other non-local areas may be easily received. There are differences too, with radio and television reception, so this knowledge is crucial to utilities and others in preparing critical information to be disseminated to the public via this route. Deadline times for the transmission of news have to be met. This requires discipline by those managing the incident.

The golden rule must be, *factual and timely communication: keep it simple.* This is often difficult to achieve in practice. There is no doubt whatsoever that articles, statements and headlines that are published in the media, influence opinion, are generally believed and considered by the public to be the main source of definitive and accurate information. If the messages are negative or nondescript, much reputational damage can ensue. The time taken to rebuild public confidence can be considerable after such episodes.

The former South West Water Authority were heavily criticised for the lack of accurate, meaningful, timely and sometimes confusing information given to the public during the early stages of the Lowermoor Incident in July 1988. The 'Clayton Report'[9] acknowledges this by agreeing that the incident, its subsequent adverse publicity and promotion of inaccurate information combined to cause serious anxiety to some vulnerable individuals who "will be slow to recover".

Some of the media reporting of this incident was inaccurate and without scientific foundation but subsequent reports became balanced and detailed. Statements such as:

(i) 'On 6 July 1998, 20,000 people in a remote part of North Cornwall ... were poisoned' - Television South West, November 1988.

(ii) 'South West Water Authority had attempted to cover up the dumping for six weeks, during which time 20,000 people became ill' - Observer, December 1988.

(iii)'(South West Water) ... is still supplying its customers with water that poses a serious threat to their health' - Sunday Times, November 1988.

(iv) 'People in Cornwall are crying out to be examined' - BBC 'Nature' programme, February 1989.

caused unnecessary suffering and anxiety in the opinion of the Lowermoor Incident Health Advisory Group who commented that –

'statements (i) and (ii) were at odds with the 'records of local general practitioners who reported no increase in consultation rates. Similarly, this Group was not aware of any adequate scientific information to support statement (iii), whereas specialist medical advice was always available to the affected community via the general practitioner service'.

The headline – 'History repeats itself in the Worcester drinking water incident' - appeared in the ENDS Report 235 in August 1994 and refers to pollution of the River Severn in April 1994. A sub-heading in this article has the title - 'Forgotten lessons of the Dee incident'. In this paragraph, reference is made to - "the provision of public information ... was prompt and effective, albeit with some local shortcomings."

Water utilities and public health organisations will continue to change their organisational structure, their culture and personnel, hence the need to remain focussed and accurate in continuing to develop and maintain vitally important communication methods and pathways. In the recent past, systems of work have been formalised, with procedures for a wide range of functions and tasks being quality assured. During emergency situations affecting water quality and sufficiency can we be confident that communications will be adjudged to be of Assured Quality. It must be the aim of water utilities to ensure that for the future their communication systems are like their other systems of work – Quality Assured.

7 ACKNOWLEDGEMENTS

The views expressed are personal and do not represent the policies and aims of any of the organisations to whom reference has been made.

References

1 J. Thurber, "Friends, Romans Countrymen, Lend Me Your Ear Muffs" - Lanterns and Lances, 1961
2 C.S. Short, J IWEM, 1988, 2 August.
3 Report of the Director of Operations (Yorkshire Water) on The Contamination of Boreholes at Bramham (North Leeds) during July 1980 - August 1980.
4 Joint Report on the Pollution Incident Affecting the River Dee and Water Supplies. (Welsh Water, North West Water). January 1984.
5 Report of Dee Working Party on Actions Required and Taken Following the Pollution Incident Affecting the River Dee and Water Supplies. January 1984 - July 1985.
6 Actions to Minimise the Effect of Pollution Incidents Affecting River Intakes for Public Water Supplies. Water Authorities/Water Companies Associations - June 1984.
7 An assessment of the Contribution of Scientific Services - South West Water Authority, into the Investigation and Resolution of the "Lowermoor Incident" - July 1988 - C. Brian Buckley. November 1988.
8 Report of an Inquiry into an Incident at Lowermoor Water Treatment Works of South West Water Authority on July 1988 - Dr. John Lawrence - August 1988.
9 Water Pollution at Lowermoor, North Cornwall - First/Second Reports of the Lowermoor Incident Health Advisory Group, July 1989/November 1991.
10 C.B. Buckley, JCIWEM, 1997, 11 April.
11 Water Undertakers (Information) Direction 1998 - Drinking Water Inspectorate. Information Letter 4/98
12 Operational Guidelines for the Protection of Drinking Water Supplies - Safeguards in the Operation and Management of Public Water Supplies in England and Wales - Water Authorities Association, September 1988.
13 Principles of Water Supply Hygiene. Water Services Association/Water Companies Association, June 1996.
14 J. Badenoch , *Cryptosporidium* in Water Supplies, HMSO, 1989
15 Guidance on Safeguarding the Quality of Public Water Supplies. Dept. of the Environment/Welsh Office, 1989
16 C.B. Buckley, Proceedings AWWA Annual Conference, New York - June 1994.
17 Environment Agency Wales (personal communication) January 2003.

TESTING THE SYSTEM – EXERCISE IS GOOD FOR YOU !!

W.T. Sutton

Yorkshire Water Services Ltd, Western House, Western Way, Halifax Road, Bradford, BD6 2LZ

1 INTRODUCTION

The basic principles of testing or exercising any system which involves the monitoring of products or activities with a view to detecting the unusual and then responding appropriately are the same whatever the product of activity may be. For the purposes of this conference I will tend to confine my presentation to water contamination emergency responses but I suggest the principles I will talk about apply to all types of incident/event or emergency responses to a greater or lesser extent.

In this instance, it is to consider the contamination of water at any point in the process of collection, treatment, or transmission to our customers and our systems for detecting, identifying, and dealing with that contaminated water in the most appropriate way, so that we may resume normal service to our customers as soon as possible. Nor must we forget how we provide safe alternative sources of water for our customers whilst the contaminated water continues to pose a risk to their health. What then are we to test? My view is the whole process for if we miss out any section we leave ourselves unsure as to how effective our systems and procedures are.

2 COLLECTION POINT

Our first real opportunity to detect contamination in the water is at the collection point, be that in an impounding reservoir, a river, or aquifer. The first question is do you have the means of detection in place and what are the limitations on its capability: –

- Is it maintained, calibrated, working?
- What is it capable of detecting?
- What is your notification system for detection?
- When was each aspect last tested and if not, how do you know it is working now?

To achieve the maximum value from detection it is necessary to have early notification and a plan of action which will:-

- Prevent the unknown contaminant progressing into treatment or supply until we know what it is and its effects

- Identify what our options are to continue to provide water to the customers usually provided from that source – where from and for how long it can be sustained
- Identify the need for delivering water by alternative means e.g. tanks/bowsers and bottled water
- Identify vulnerable customers so that special arrangements may be made as necessary
- Indicate what, if anything, we need to be telling our customers and how we are going to deliver the message
- Identify which laboratory/ies samples will be sent to and what we will be asking them to look for
- Include information and intelligence which will inform our decisions
- Ensure Regulators and other agencies are notified
- Identify options for disposal of the contaminated water and timescales
- Provide a regime for testing the system – when was all of the above last exercised?

3 TREATMENT

A water treatment works provides a number of opportunities for contamination both accidental and deliberate. Deliberate contamination is really a security issue and water treatment works require high security protection in terms of their physical and electronic security. Security systems physical, procedural, and electronic also need testing and responses need exercising – when did you last carry out a penetration test on your treatment sites – or indeed when did you last walk the site and conduct a security audit?

Accidental contamination should not occur if good quality site procedures and appropriate facilities are in place – but how often are those procedures subject of live exercise, or tasks covered by them observed to ensure compliance and minimise risk. I know of one instance in respect of deliveries of chlorine drums which required the delivery vehicle to be driven into the store room for offloading and the storeroom door to be closed whilst that took place. When the activity was observed the vehicle was found to be too large and made the closure of the door impossible. That vehicle had been used for the delivery for several months. How often do your managers find time to monitor activity on sites or run through normal business procedures with their teams?

4 THE DISTRIBUTION SYSTEMS

Once again physical and electronic security systems provide a significant degree of protection at the main vulnerable points and the way in which systems operate provides further elements that assist in securing supplies. It is not appropriate in this paper to disclose the nature of those vulnerabilities nor indeed the means of protecting them. However it is relevant to prompt an examination of those vulnerabilities and protective measures and to suggest that periodically they are tested to evaluate their effectiveness. It is of course essential that everyone is clear on the notifications and actions they must undertake in circumstances that suggest security has been breached:-

- to ensure that the water within the system can be isolated from supply to the customer
- restricting the flow of water already in distribution (option of dropping the pressure)

- to arrange immediate sampling - use of field kits if available as a first screen
- gathering information from the scene – evidence for the police
- liaison with Regulators and Health Authorities
- liaison with the Police
- identifying the area and customers which are being provided by the system and whether there is an alternative piped supply available to some or all
- issuing advice to customers
- identifying how far the water may have travelled from the point of suspected security breach
- options for the disposal of any contaminated water
- means of cleaning out the contaminant and disinfecting the system.

Exercising the procedures will identify the weaknesses in the systems and procedures, allow you to practise in a secure learning environment, and give you the opportunity to take corrective/preventive action.

5 ANALYSIS

I have dealt mainly with testing the water company systems or procedures but many of the actions and options are going to be dependant upon the ability of the laboratory services to come up with fast accurate analytical results upon which we can base decisions. It is therefore essential that laboratories have systems in place for receiving large volumes of samples in a crisis and undertaking analysis for 'unknowns' as opposed to the 'usual suspects'.

The UK Water industry and our laboratories have, over the past 5 years, carried out a series of exercises in which up to 10 laboratories have simultaneously undertaken the analysis of controlled spiked water samples. The purpose was to establish the contents, and estimate the approximate volume that had been introduced into a known quantity of water plus a comment on toxicity. The results and methodologies employed have been shared by the laboratories and have formed the basis of an approach to analysis of such mystery samples. Such exercises are worthwhile in identifying gaps in technical knowledge or equipment deficiencies. Some may argue that they also provide an indication of the loss of those scientists who have been well practised at thinking and working through such problems but are being replaced with people who are very good at feeding the sausage machine of routine analysis upon which profitability depends.

The bean counters will always support the principles of Fordism and division of labour but take much more convincing about the need for retaining the services of those highly qualified, skilled and experienced people who may be considered over qualified for 95% of the work received in to the laboratory. But when the dirt hits the fan there are going to be very few 'grey beards' available to lead the analysis process from first principles. It is at such times that collective knowledge and networks amongst scientists can become a powerful positive influence. It is such concerns that have led to increasing co-operation across the industry and mutual aid arrangements amongst competitors as a means of providing resilience both to their own business and to the water industry as a whole. So far the arrangements have not been tested in exercise or for a real emergency. The nearest thing so far provided Thames Water with the opportunity to demonstrate the robustness of their Business Continuity Plans when they coped remarkably well to the bombing of their laboratory at Canary Wharf by the IRA. How many other laboratories have similar plans – and when were they last tested?

Progress has been made in terms of producing records of the capabilities of each laboratories and 24 hour contact arrangements. However, if they and we as an industry are to be able to cope with a large scale emergency requiring large volumes of samples to be tested for other than routine regulatory purposes there is more to be done. In particular in respect of standardising of records, containers and equipment specifications so that we have consistency wherever a sample may be despatched. Simple exercises would quickly identify those issues and provide clear guidance for the future. Such exercises need to be extended to those Government Laboratories which we may call upon through the Drinking Water Inspectorate in an emergency so that once again we may know just what sort of volumes of samples can be dealt with and what is the turn round time from receipt of the sample to an analysis result which says the water is safe to drink.

6 COMMUNICATION

In every major emergency, or exercise, communication has proved to be one of the most significant elements – if you have got it right you are believed to have done a superb job; if you do a magnificent job but get the communication wrong it will be perceived that you got it badly wrong.

We are in an industry which relies heavily on the trust which the consumer has placed in us, but trust is won the hard way and over a long time – it is also lost very quickly and takes even greater effort to win back. A major contamination event will not just have an impact upon the individual company or companies affected but across the whole industry, possibly world-wide. We therefore all need to be prepared to communicate with our customers as either a direct or indirect victim. In all crisis communication clarity will be of the essence – defining which area is affected down to street level, what the cause of the problems is, when it is likely to be resolved and what alternative provision is to be in place in the meantime. That message needs to be out there as soon as possible and apart from some of the detail should be written now. We also need to be prepared for the supplementary questions because no doubt the TV pundits will be seeking full details of sampling regimes and demanding re-assurance that the glass of water just drawn from a studio tap is safe to drink.

That scrutiny will not just be restricted to the industry but will also include our Regulators both in terms of quality and financial arrangements. It is the Regulators that the people, through the Government, have appointed to keep the industry in order. If we fail then some may view it that the Regulators and Government have also failed. Our responses to such a line of questioning needs to be jointly considered and rehearsed now, because if the responses you give in exercise today are unconvincing just think what they are going to sound like when an aggressive investigative journalist is asking them live on Newsnight.

I will not be sitting in the press conference or TV studio answering those questions it will be one of our Directors, or the Drinking Water Inspector, or the Director General of OFWAT, or the Health Professionals, or the Minister responsible for Water. When did they last have a joint media response exercise around such a scenario, and what guidance has already been produced for them?

7 TEAMS AND PRACTICE MATCHES

Just as you would not expect any professional sports team to turn up for their match without practising together so we should not expect ourselves to be able to respond as a cohesive team in an emergency without playing a few practise matches. Remember this, the sports teams are playing for honour or financial reward – you may well be playing for the life and certainly the health of your customers and your company. You have to test your team, see where there is flexibility, identify strengths, work on the weaknesses, consider the strengths and weaknesses of the opposition, pick your best team and best structure, and then play together.

The great golfer Gary Player was once accused of being a lucky player and he responded "You know, I have found that the more I practice the luckier I get."

Exercise is good for all the team whether you be the captain or one of those on the reserves bench. I also think this conference emphasises that we are a team because it includes many elements, not just from the water companies, not just the scientists, but also the Regulators, health professionals, the laboratories and academia and all with one purpose – the provision of potable water to our customers – the public.

So please get your practise in now. Test OUR systems before the emergency. Mistakes made in exercise are far less costly than those during a real event. You do not want a major emergency to be the first time you have tested your systems.

Exercise is good for everyone.

CAN WE COPE?

P. Mills
Water UK

1 INTRODUCTION

I am going to focus on drinking water rather than the environmental aspects. And I am not going to cover identification of contamination and contaminants. That real issue was covered by others yesterday.

So this is the denouement. Decision time. The answer to the fundamental question, like the meaning of life, and the universe. So from everything you have heard these three days and from what I am about to tell you now, - *"Can we cope?"*

Let us consider what we really mean by that. I was not party to the conference planning process but can infer what the committee had in mind. It is a common enough word - to cope. An alternative would be to "manage", or to "get by"

But more importantly, what are our customers' expectations? I believe it is getting a high quality piped water supply 24 hours a day, 7 days a week. And the industry has a very good record of delivering that. "Always on tap" is a common expression – and its meaning today extends far beyond the original water context – and that is significant. And that expression influences the 'can we cope' question.

So what do customers expect in times of emergency? And that means any emergency. We have heard about natural and man made incidents covering chemical, biological and nuclear or radiological contamination. But now we also have to consider the real threat of terrorism and deliberate and hoax contaminations

So I am going to cover
- Background
  from a UK perspective, where we are and where we have come from
- Basis
  the principles and strategy that we adopt – including the assessment of risk
- Arrangements
  highlighting the arrangements the industry has put in place to protect our water supply
- Response
  considering the capability of the industry to respond to events and provide alternative supplies

Finally I'll recap the significant points and the measures we have taken to cope with a water contamination emergency

2 BACKGROUND

First of all let us consider the context of the water industry. It will help understand our customer expectations.

- It is a provider of essential and public services
- It is a key contributor to the health of the nation
- The quality of drinking water is at a record high – 99.86% of 2.8 million tests complied with the regulations in 2001 (the latest reported year), and
- It is a guardian of the environment and the biggest UK investor in environmental protection

Yet so often it is just "taken for granted." So it is not surprising that customers have not really thought about water not being available. So can we cope does not really mean can we meet customers' expectations - since their expectations are based on maintaining the existing arrangements – a high quality 24/7 piped water supply.

Now let us consider the position in the UK. And if we are talking about a terrorist threat to the public water supply we are speaking from a very different position to say the US. And that is important. Following 11 September and the natural rush to protect the US Homeland we were almost inundated with calls from the media. They are doing this and that in the States. What are you doing? Why aren't there big stories and lots of activity in the UK? Two reasons:

a) For mainland Britain, unlike the US, security threats had been real, credible and actual since the 1970s. So the need to protect essential assets and services here was not new

b) Security arrangements, to be effective, need to be secure. Government policy, supported by the industry and its Regulators, is not to discuss these arrangements. But inevitably there is a balance. A balance between remaining tight-lipped and secure and thereby forcing people to make their own (probably wrong) conclusions, or saying enough to reassure and give confidence to the public, through the media, that effective measures are in place. And this conference, this presentation, is another opportunity for reassurance without compromise.

So companies have been aware of the threat from the provisional IRA and have responded with the help of government to that threat improving the protection and resilience of key installations over many years. The security threat in the UK and the nation's and the industry's response predates September 11 by about 30 years.

But there can be no complacency. The 11th September and the Al Qaeda threat has raised the stakes. There is a new awareness and the water industry has increased vigilance, reviewed and revised plans, increased investment and has ongoing discussions with government. We have fast-tracked some measures, considered new scenarios and developed new guidelines and practices, including for example developing a joint protocol with Environment Agency and DWI primarily for the emergency services covering the disposal of water contaminated from treating casualties. This followed on from the anthrax scares of 2001 and 2002. We have also developed an industry guidance document on managing access to the pressurised water network.

Despite the awareness and increased vigilance the threat level advised by Government has not changed because of September 11. There has been no specific, credible threat to the water infrastructure in the UK. Obviously any action in Iraq may affect that security level. And if there are changes the water industry will respond to the guidance from government.

3   BASIS OF PREPARATIONS

Having covered the background, let us move on to the basis of our preparations – the principles and the strategies that the industry in conjunction with government adopts – to protect both customers and the national interest.

Not unsurprisingly we adopt the principle of risk assessment and management. Risk can be assessed from the probability of an incident occurring (such as an attack on the system or the failure of equipment) and the consequential impact of that incident. I will come back to this.

So our strategy is based on risk assessment of the water supply infrastructure and assets, that is - the WTWs, pumping stations, reservoirs and trunk mains. Elsewhere this may be known as a vulnerability assessment. And to be able to respond effectively to threats we have to consider the "worst case scenario" – traditionally the incident having the greatest impact population wise, say a WTWs being out of commission, a burst on a strategic trunk main, a contaminated reservoir – from whatever cause. However current thinking, adopting a true risk management approach considering both probability and impact, suggests the incident with the greatest risk should be the worst case scenario.

And to take account of terrorist threats we have to consider multiple, simultaneous incidents not just a single worst-case scenario.

But as well as responding we have to protect the infrastructure and assets. We need to know the probability of threats to the system. The government security advisers provide us with this intelligence. They assess the threat or alert level and advise whether threats are credible or credible and specific. If the threat level changes the industry would implement additional measures, on top of the existing security protection.

The basis of our response capability is specified in the Securities and Emergencies Direction. This requires companies to have Plans in place, including resources and equipment to deal with worst-case scenarios and Plans in place for multiple simultaneous incidents. It was issued in 1998 and requires company to review these plans and have them independently audited and reported to DEFRA each year.

Summarising our strategy and response to risk. If we consider risk as the product of the probability of an attack and the impact of that attack we can minimise the risk in two ways:

    a) By minimising the probability - by providing protection and increased security measures to our sites and increasing their resilience

    b) By minimising the impact - by having an effective capability to respond and recover the situation and/or provide alternative supplies.

4   ARRANGEMENTS

So on that basis and with that strategy what arrangements are in place? Obviously I can only give an overview here

To secure and protect the essential infrastructure requires a partnership between government and the industry. Government sets the framework for measures with the SEMD (???) and companies develop the plans and have in place the resources and equipment or have access to resources and equipment to deliver those plans

Government also collects and communicates intelligence to the water industry on potential threats and advises on any changes in threat level. The industry responds to the threat level according to predetermined escalation procedures.

In addition to developing their own Plans to meet the SEMD requirements the industry, through Water UK and its Emergency Planning group, has developed, and keeps up to date, an Emergency Manual and several codes of practice. Some of these are recent. I mentioned the protocol for disposal of contaminated water developed last year in response to the anthrax threat. And the code of practice for the control of access to the pressurised water network. Other codes have been around for some time, such as the code on protection of service reservoirs. It categorises and specifies the security arrangements that companies must put in place at their service reservoirs. This code was revised this year.

And as you would expect, there is significant and focused capital and operational investment - whether on WTWs, service reservoir sites, or reviewing Plans and providing emergency drinking water bowsers and static tanks.

A further significant contribution to the protection and resilience of the system includes the arrangements through DWI with external laboratories for water quality analysis beyond the remit of water company own laboratories.

There are also clear communication channels and arrangements within the industry from government to the companies and between companies themselves to share information and intelligence.

And last but certainly not least are the necessary exercises carried out by companies - both individually and nationally - to test the system. As Bill pointed out this morning – it is only by probing the system that we can learn and strengthen our resilience.

5   RESPONSE

The other side of the coin to security and protection is having an effective capability to respond .The two are complementary and together by minimising the probability and by minimising the impact they reduce the risk of disruption.

The responsibility for responding is clearly down to the industry. But how? The industry produces and revises annually their Contingency Plans to meet the worst-case scenario. The requirements for these are set down in the SEMD.

To deliver these plans the industry has invested, and continues to invest, in alternative means of supply. There are around 12,000 water bowsers and static tanks held in store by the industry that can be deployed if the mains supply is no longer available. But it is not practical and not cost effective for every company to keep in store and ready for use all the equipment it could possibly need for multiple simultaneous scenarios. The industry therefore developed a system of Mutual Aid, where resources and equipment are shared between companies in times of need. One of the workshops on Monday evening discussed this – it is an essential component of our resilience. The Wem incident, describe on Monday, demonstrated its critical role. But obviously it is essential that we maintain an accurate and up to date record of equipment and its location. And this is one of the roles of the Water UK Emergency Planning group

I mentioned lab support previously. It is sufficient to say that arrangements are in place on a 24/7 basis to support company labs analysing and identifying materials not normally present in raw or treated water supplies.

There are channels set up within the industry to communicate intelligence from government and also share information between companies. Part of that communication is notifying changes in threat or alert levels. And when an alert level changes the industry must respond and implement specified measures to further increase security and vigilance.

There are also local communications links in place with the health sector through Incident Management Teams and EHOs.

In times of crisis there also needs to be timely, clear and helpful advice to customers. This not only has to be in written form but also through the local media. It is essential to keep them involved and 'on-side' - they have a key role to play in shaping customers' perceptions and helping manage their expectations. There will be a demand for information. Not providing this is not an option.

The final part of this equation is the provision of alternative supplies. The industry has a statutory duty to supply water even if it is not by the conventional and accepted means. I have already mentioned bowsers and static tanks. But for vulnerable customers companies normally provide emergency bottled supplies. Some companies tend to use bottled supplies rather than deploy bowsers or tanks if the area or number of customers affected is small

But in order to plan responses and manage the deployment of alternative supplies its first key to know – How? And how much?

This single factor has a major bearing on the industry's strategy and initial capability to respond.

But again - what are our customers' expectations?

I cited the "taken for granted" assumption and "always on tap" expression. So it is a shock when water is not "on tap." Collecting water from tankers on street corners does not meet customers' expectations – but it is providing an alternative and short-term supply.

So "the how" is essentially tankers topping up local 1100 or 1800 litre bowsers and static tanks or delivering bottled water to every household. But both options have significant storage, management and logistical issues. And some factors are outside the industry's control. The logistics are dependent on external transport contractors coming up with sufficient vehicles at less than 24 hours notice. A further issue is some people hoarding supplies or taking supplies meant for others.

But perhaps more critical is the volume of water that has to be delivered. In the UK the Notification to SEMD states not less than 10 litres per person per day – Why?

We all know that we should be drinking 2 litres or thereabouts of water a day (though how many of us actually do?) So where is the other 8 litres going? – especially if there is still water available to flush toilets. And how many people do we assume are in each household, 2, 3, 4 – it makes a difference!

In the US the Federal Emergency Management Agency recently published its guide to citizens' preparedness. It recommends (although in a slightly different context) an emergency requirement of 1 US gallon – i.e. just under 4 litres per person per day.

Do customers really expect 15 2-litre bottles on their doorstep or will 6 do? Would they carry 30 Kilograms of water back from the bowser or just 6 kilos? Let us say it is an issue to resolve – especially for that first 24-hour period.

It all boils down to finding a real balance between the logistical practicabilities and managing and meeting customers' expectations.

6   CONCLUSIONS

So what conclusions can we draw? As, I have explained it is all about minimising risk. The industry minimises the risk of water contamination emergencies – from whatever cause:

- By carrying out risk or vulnerability assessments of its installations and networks
- By providing security measures and protecting those installations
- By having response plans in place that are regularly reviewed and tested and having the resources and equipment in place to deliver those plans

And

- By having good communications and sharing intelligence to and from government and within the industry

So "can we cope?" – can we manage, can we get by.....?

Well we have all the measures in place, tested and ready.

SUMMARY OF WORKSHOP OUTPUTS

M. Furness

Quality and Environmental Assurance, Severn Trent Water

1 INTRODUCTION

On the evening of the first day of the conference, four workshops were held. Delegates could choose which one to attend. The topic for each was aligned to the main objective for the conference –Water emergencies – can we cope? Could we identify subject areas or activities where we felt lacking in capability? To provide a focus for each workshop the following topics were chosen:

- Analytical methodologies
- External communications
- Mutual aid – operational support
- Public health and toxicology

A member of the organising committee chaired each session, drawing on experts in the audience where necessary, to challenge assertions and views expressed. The delegates were encouraged to share experiences, best practices and any concerns within each topic area. Finally, a number of recommended actions were identified from the outputs for the water industry to consider.

2 ANALYSIS ISSUES

This workshop was led by Steve Scott of Thames Water and Clive Thompson of ALcontrol Laboratories. The primary areas of discussion were:

- Resource availability and competency recognition;
- Analytical capability;
- Quality control;
- Communication on laboratory expertise areas; and
- Funding and associated risks

Most analytical services have now centred on a small number of high throughput laboratories that greatly utilise automation. This is based upon both low- and high-tech methodologies with associated robotics. These large laboratories are very efficient at carrying out the targeted analysis of very large numbers of samples. The demand for highly skilled analytical chemists has declined over recent years leading to a potentially serious gap with respect to interpretative skills, knowledge sharing, and technical intuition needed for rapid non-target interpretive analysis of emergency water samples that could be chemically, radiologically or microbiologically contaminated. Allied to this, the group felt that the status of Chartered Chemist and Chartered Biologist is not promoted to the same extent within the water industry as in other industries and is less valued by analysts and microbiologists. With the normal business pressures now vigorously applied to laboratories, staff exhaustion can occur during prolonged incidents when 24-hour working may be expected. Throughout the discussion, there were several references to manpower concerns raised by individual laboratories. Certain topics (e.g. the better recognition of C.Chem and C.Biol status) would be best tackled through one voice representing all laboratories.

In terms of technical/analytical capability there were a number of issues raised. It was felt that the industry should have more robust broadband screening tests and particularly those methodologies where there was universal agreement on their detection capabilities. Primary screening for a wide range of pollutants needed improvements in terms of breadth, sensitivity and timeliness of result. To complement laboratory testing there should be improvements in on-line monitoring capabilities. It was felt that pathogen analysis in the water industry was too limited and the Health Protection Agency (HPA) service may not be able to respond effectively to major local incidents. More interlaboratory comparability studies were recommended for potential pathogens. It was noted that the CSL LEAP® "Mystery" proficiency scheme only provides chemical determinands. Of particular concern was the capacity of a few specialist laboratories to undertake radiological analysis of large numbers of samples. This was particularly apparent during the Chernobyl incident.

The ensuing discussion culminated in a call for greater knowledge sharing, awareness of best practice and designated centres of excellence for identification of unknown agents in water emergencies. Funding should be made available for common research into incident analysis. At the very least there should be a database accessible to all relevant staff to aid interpretation of unknowns.

All laboratories should follow set protocols for analysis of unknowns. One theme, common to all the workshops arose at the end of the discussion. As an industry, we should share information on incidents, hoaxes, threats etc., and derive lessons to be learnt for all. This has been partially achieved by the setting up of a Mutual Aid Group for laboratories from the UK water companies and other key laboratories. This "self-help" group was formed eight years ago and meets on an annual basis. It has set up a database of emergency response capability statements compiled by each participating laboratory with 365 day / 24 hour emergency contact numbers.

3 MUTUAL AID SUPPORT

This workshop was led by Bill Sutton of Yorkshire Water. The principal participants in this workshop were Emergency Planners, Operations Managers and Laboratory Managers. There was initial discussion on what mutual aid means to the water industry. Very much led by Emergency Planners from all the major water companies, mutual aid was originally a forum which was developed to provide an operational support mechanism for responding to water emergencies and includes provision for tankering, bowsers, generators, pumps etc.. A much wider organisational involvement can now be called upon including local authorities, other utilities and emergency services should they be needed.   More recently, a support forum for laboratories has been established. Capability data has been shared and progress is in hand to share more widely expertise during emergencies, without conflicting with commercial interests.

The value of mutual aid for water emergencies was discussed and reinforced:

- Increased national resilience;
- Intellectual sharing and interaction;
- Increased confidence in ability to cope; and
- Reduction in capital costs.

There were however other issues to address including:

- National standards for equipment/procedures;
- Clarification of commitments with respect to deployment and costs; and
- Cost reimbursement.

There was strong support for a mutual aid forum continuing. During 2002/3 there had been collective industry support during the response to two water emergencies. Mutual aid and support was felt to be cost-effective. It increased resilience and is at the core of integrated emergency planning.

4 COMMUNICATIONS

This sessions was facilitated by Michael J Scott of Sensors for Water Interest Group (SWIG). Following initial prompts, a general discussion on the current state of Water Company communication cascades ensued. It was felt that for most water companies internal communication was effective and appropriate. Inter-company liaison could be improved though particularly in sharing lessons learnt.   The value of proactive information sharing for likely events was supported. There was then a brainstorming session on factors influencing communication flow:

- Could financial considerations have undue influence on risk assessment?
- Do glass ceilings exist?
- People do not always respond as expected
- Can customers become complacent?

Some important principles were established:

- Communication is always multi-faceted;
- There is a need to inform early;
- Scale up the response before the problem escalates, it can always be scaled down more easily;
- Keep up the tempo, do not have a news vacuum;
- Target the media deadlines;
- Real experience does not come from exercises;
- One team, one message; and
- Accountabilities must be clear.

Finally a number of points were highlighted to reinforce the practice of good communication. It was recognised that high mobility and skills depletion can present modern day challenges, but can and must be overcome. The workshop finished with a key enabler to good communication with external organisations - "Foster relationships – both formal and informal, with external bodies".

5 HEALTH RISKS, TOXICOLOGICAL ADVICE AND BIOHAZARD ISSUES

This workshop was led by Martin Furness of Severn Trent Water and John Fawell, Consultant in Drinking Water and Environment. Although the conference theme avoided the topic of wilful use of biohazards, it was felt appropriate to allow some discussion in this workshop.

The discussion started with thoughts on databases. A number of organisations draw on different databases for health risk assessments. The Water Industry, through an UKWIR involvement has its own TOXIC database. Greater expert collaboration was called for to promote consistent interpretation. Allied to this, the value of medically based assessments as used to establish World Health Organisation (WHO) standards was reinforced. A topic not receiving much exposure at the moment is that of disposal of any contaminated waters. The group felt that more should be done with regulatory agencies, emergency services and water companies to define sensible procedures for credible scenarios. The question of how we should analyse for novel toxic substances drew a wide-ranging response. More reassurances need to be sought that we in the Water Industry have the best techniques available.

We must promote expertise sharing or develop one or two specialist centres of excellence. This work would need to be allied to services provided by the Health Protection Agency (HPA) and the services at DSTL and LGC. Further research on novel detection systems and in particular bio-monitors was encouraged.

A critical step in water emergencies is risk assessment. What impact could there be on public health and what level of protection should be in place? Wider engagement of public health professionals to promote greater degrees of reassurance to customers was advocated. Stronger liaison with the HPA and local HP teams, particularly over background surveillance was thought sensible. More high-level liaison with HPA was called for. For the future, alignment with the hazard analysis critical control points (HACCP) work to come was put forward.

6 SUMMARY OF ACTIONS TO BE TAKEN FORWARD

- Concern about recognition and value of qualified scientists to be raised through RSC and other appropriate bodies.

- Better standardisation of existing analytical methods and development a range of robust rapid screening methods needs to be taken forward through the Standing Committee of Analysts (SCA)

- Enhance detection capabilities to be developed through the UK water companies and other key laboratories mutual aid group. This information to be shared with all users.

- Bio-monitoring at on-site testing sites and on-line testing to be explored through Sensors for Water Industry Group (SWIG).

- Outputs to be discussed with Water UK and the water industry with collective actions for key participants, notably high level liaison with the Department of the Environment, Food and Rural Affairs (DEFRA), the Drinking Water Inspectorate (DWI) and the Department of Health (DoH).

- A forum to be sought whereby water industry incidents may be reviewed and best practice shared on operational responsiveness, communication cascades and water quality monitoring.

- A public health liaison network to be established with the Health Protection Agency (HPA) and involving the Chartered Institute of Environmental Health, DoH and DEFRA.

Subject Index